SpringerBriefs in Energy

Markus Lehner · Robert Tichler
Horst Steinmüller · Markus Koppe

Power-to-Gas: Technology and Business Models

 Springer

Markus Lehner
Industrial Environmental Protection
Montanuniversität Leoben
Leoben
Austria

Robert Tichler
Horst Steinmüller
Energy Institute
Johannes Kepler University
Linz
Austria

Markus Koppe
Institute for Environmental Management
Johannes Kepler University
Linz
Austria

ISSN 2191-5520   ISSN 2191-5539   (electronic)
ISBN 978-3-319-03994-7   ISBN 978-3-319-03995-4   (eBook)
DOI 10.1007/978-3-319-03995-4

Library of Congress Control Number: 2014943943

Springer Cham Heidelberg New York Dordrecht London

Printed on acid-free paper

Springer is part of Springer Science+Business Media (www.springer.com)

Preface

The change in the supply structure for energy is mainly driven by the imminent climate change. Other incentives may be strategic considerations, or generally a paradigm shift in the way our industrial system, and the necessary power supply is operated. The energy supply of the future will implement renewable sources at least to a greater extent as today. Beyond any controversy, increasing portions of renewable energy, particularly wind and solar power, already cause local discrepancies between supply and demand in the power grid.

There are several possibilities to approach the challenges of a changing energy system. For the time being, the extension of the power grid, load management and energy storage facilities are possible measures to meet the requirements of renewable energies. Depending on the future rate of renewable energies, most or even all of these measures have to be implemented. In terms of storage systems, also seasonal storage possibilities are needed. One promising option for long-term storage is the conversion of renewable electricity to chemical energy carriers, like hydrogen, methane, methanol, formic acid, fuels or the hydrogenation of aromatic hydrocarbons.

The intention of this book is to give a brief, but comprehensive overview of the Power-to-Gas technology, one of the chemical storage options for renewable energies. Many researcher groups are currently working on different aspects of this concept. Power-to-Gas plants in a demonstration scale were recently started or are under construction. Therefore, it is not possible to give a concluding résumé of this technology at present. Furthermore, the Power-to-Gas concept is a flexible technology providing a multitude of possible applications. In order to cope with this situation, we tried to describe the current state of the art, actual research and development activities as well as future challenges, without making a claim to be complete. The second part of this book deals with business models focusing on the economic dimension of the Power-to-Gas technology respectively of the Power-to-Gas system, which requires not only business analysis but also comprehensive macroeconomic and systemic analysis.

Currently, the Power-to-Gas technology is economically not feasible. Both, still technological and systemic developments are required. But, in the opinion of the authors, the long-term storage of renewable energies will be a crucial backbone of the future energy system. If we do not develop technologies today, we will not be able to meet the requirements of tomorrow.

The authors would like to thank Dipl.-Ing. Aaron Felder, Dipl.-Ing. Phillip Biegger, Prof. Dr. Josef Draxler, Lukas Rebhandl, and Fabian Frank for reviewing parts of the manuscript, and Mark Read as well as Jed Cohen, M.S. for transforming and partly translating the text to a readable English.

Leoben, May 2014   Markus Lehner
Linz   Robert Tichler
                                                                              Markus Koppe
                                                                          Horst Steinmüller

Contents

Chapter 1
Storage Options for Renewable Energy

In the recent years, the European energy policy has agreed on the increased integration of renewable energy sources in the energy system, and large efforts are being made to implement renewable energy. This tendency is not limited to the European market, but is a basic development in many regions. The energy policy is primarily based on climate change policy aims and demands, however further parameters are relevant in the portfolio of intentions for increasing the percentage of renewable energy sources, such as reduction of the import dependency and increasing the domestic value or price stability. To some extent, relatively high expansion rates in the implementation of energy systems based on renewable sources can be achieved, such as in Germany and China, for example.

The increasing share of renewable energy sources, in most cases coupled with an absolute increase of production, includes as well as advantages, challenges and problems. With this in mind, this book concentrates on the challenges of a continuous increase in the volatile portion of energy production caused by renewable energy sources.

Renewable energy sources are being forced into all areas of energy systems: in the area of mobility with respect to fuel, thermal area (both as energy sources in the segment of space heating as well as in process heat) and in electricity. This book focuses on the challenges in the area of electricity production. The areas of heat and mobility are therefore in this respect not relevant for the problem of necessity of storage systems for volatile production lines (certainly however as a demand in the Power-to-Gas plant produced energy as detailed in Chaps. 2 and 5). As a consequence, only the challenges of volatile power production on the basis of renewable energy resources will be dealt with.

The continuous increase of the volatile portion of power production based on the energy policy road maps is not equal in all regions. Renewable energy sources for the production of electricity such as water power or biomass show, in comparison to wind energy or photovoltaic, less temporal fluctuation in the production. Therefore, all regions with high or strongly increasing shares of wind and solar power in their electricity production portfolio are or will be confronted

© The Author(s) 2014

M. Lehner et al., *Power-to-Gas: Technology and Business Models*,

SpringerBriefs in Energy, DOI 10.1007/978-3-319-03995-4_1

Fig. 1.1 Measures to cope
with higher shares of volatile
renewable power in the
energy system

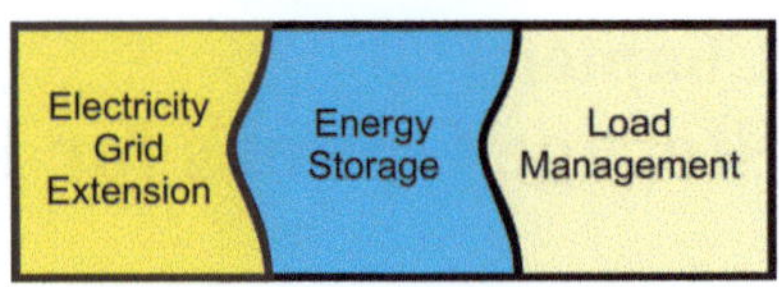

with high portions of volatile production. In the European Union, due to the national road plan, especially Germany is affected, but this pertains also to other regions such as Denmark, Great Britain or Spain with respect to the development of wind power. Spain and Italy are also concerned with respect to the integration of fluctuating solar power, respectively photovoltaic plants or the solar power stations. Due to climate conditions, a constant level of power production with wind and solar energy cannot be achieved. Therefore, the energy systems have the challenge to balance the strong fluctuations in the production. At the moment, and probably also for the next years, the problems arising from temporal and spatial fluctuating energy production are limited to special regions, and do not occur frequently. The possible measures to adjust the energy system for a high content of renewable energy sources (see Fig. 1.1) are not needed today but in the future. The time frame strongly depends on the progress of the implementation of renewable sources, but significant demands for energy storage are not expected before 2020.

In future, the electricity suppliers and producers can predict strong fluctuations in the electricity production due to increasingly better climatic prognosis systems, but this does not completely solve the problem of an intelligent integration of the production quantities. Due to primary energy efficiency, as seen from an ecological as well as an economical point of view, systems which are based on the shutdown of wind power or photovoltaic plants due to excess supply should not be pursued. A sustainable energy system integrates these production methods into the existing system. For that purpose, the electricity grid can be expanded, furthermore, different forms of load management can be applied, both in supply and demand. These solutions should be pursued and further developed (Fig. 1.1).

But, load shifting, with or without financial incentive, will not be enough to optimally integrate the volatile production into the energy system in the future. Energy storage systems will take a crucial role in the integration of renewable energy sources with volatile production structures (Fig. 1.1). Thereby, large capacities can be stored for future use—there is no more need for permanent physical adjustment of the grid.

Various electricity storage systems are currently available on the market with extremely heterogeneous development stages. These range from for several decades established technologies such as pumped hydro storage power stations for large quantities or batteries for small storage quantities, through to technologies and systems which are currently in the development stage, such as rechargeable batteries or flywheels. Electrical storage systems can be roughly divided into storage by means of mechanical energy (kinetic and potential), chemical energy (inorganic and organic) and electrical energy. The essential assessment of storage technologies is based on the analysis of relevant parameters with regard to various disciplines—technological assessment, economical assessment, systemic assessment, ecological assessment and legal assessment. Singular consideration of individual technology characteristics is too little and with regard to storage technologies, generates no optimal solution for the further development of the energy systems.

In addition, it can be stated that for specific energy systems, different applications are to be considered, and a direct comparison of individual parameters has to take into account the specific way of utilization as well as their specific system benefit. For the assessment of electrical storage systems, the following variables should be considered.

- Storage capacity
- Maximum charging/discharging power
- Possible storage duration
- Efficiency/Utilization
- System benefits
- Storage losses
- Total storage potential of all plants
- Temporary availability, guaranteed capacity (time of day, seasonal dependability)
- Investments costs
- Operational costs (resources, emissions)
- Economic impact (value added effects etc.)
- Site conditions, need for topographic intervention
- Existing infrastructure on site, i.e. power grid
- Conversion possibility, requirement for reconversion
- Public acceptance for new infrastructure projects, environmental impacts

A comprehensive evaluation of the listed variables respectively their dimensions cannot be made at this point. In Table 1.1 the efficiency (electricity to electricity), the storage capacity per plant and the possible storage time are listed, exemplary for various energy storage technologies.

Pumped hydro storage is currently the most established technology for providing control energy in the electric power system. Electricity is converted to potential energy by pumping water to higher altitudes. When electricity is needed, the water is released from the reservoir, and the potential energy is again converted to electricity by water turbines. The efficiency of this storage technology ranges between 70–85 % which is comparatively high. The installed storage capacity of pumped hydro storage varies depending on the region. But, basically, existing pumped hydro storage facilities provide limited storage capacity which will not be sufficient for higher shares of renewable energies in the future (Bajohr et al. 2011; Klaus et al. 2010). The erection of new pumped hydro storage facilities is generally difficult due to commonly low public acceptance of infrastructure projects affecting the overall appearance of the landscape.

Table 1.1 Overview of selected parameters of energy storage technologies

Technology	Efficiency	Capacity rating MW	Time scale
Pumped hydro storage	70–85 %	1–5,000	Hours—months
Li-Ion battery pack	80–90 %	0.1–50	Minutes—days
Lead acid battery	70–80 %	0.05–40	Minutes—days
Power-to-Gas[a]	30–75 %	0.01–1,000	Minutes—months
Compressed air	70–75 %	50–300	Hours—months
Vanadium redox battery	65–85 %	0.2–10	Hours—months
Sodium sulfur (NaS) battery	75–85 %	0.05–34	Seconds—hours
Nickel cadmium (NiCd) battery	65–75 %	45	Minutes—days
Flywheel	85–95 %	0.1–20	Seconds—minutes

Sources own compilation; information from (Diaz-Gonzalez et al. 2012; Beaudin et al. 2010; Chen et al. 2009, 2014)

[a] Power-to-Gas efficiency without re-converting to electricity: 50–75 %

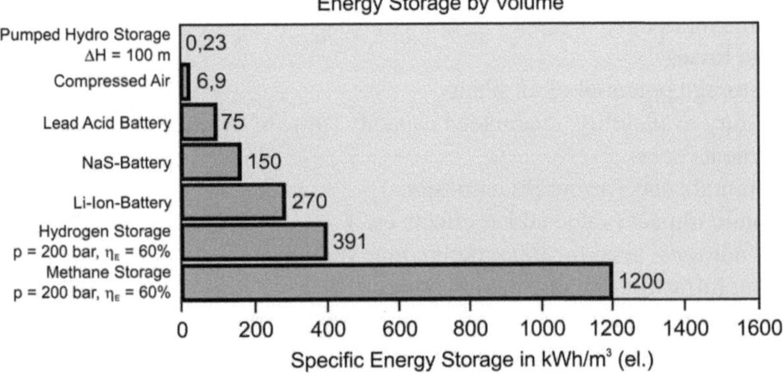

Fig. 1.2 Comparison of the volumetric storage capacity of different technologies for electricity [modified from (Bajohr et al. 2011)]

Compressed air storage converts electricity to pressurized air which is later expanded by turbines reconverting the energy to electric power again. The main drawback is the low volumetric storage capacity (Fig. 1.2) resulting in necessarily huge storage volumes. In order to achieve high efficiencies, the released heat has to be additionally utilized (Bajohr et al. 2011). The application of this storage option is additionally limited by the high costs.

Rechargeable batteries belong to the group of electrochemical storage mediums. Particularly when large amounts of energy have to be stored for a longer time, those systems show high specific costs. The gradual discharge and the degradation of the batteries limit the storage times.

Flywheels are short-term storage technologies which can absorb and release great amounts of electricity within a few seconds. But, this technology is inappropriate for long-term storage.

With regard to the storage of large energy quantities, to be stored partly for a long period of time (days to months), and which are subject to a strong dynamic emergence, the following parameters are of essential importance: high storage capacities, high volumetric storage density, system benefits, flexible site-specific modifiability, decentralized application possibility and the possible storage duration. These parameters can be well covered by chemical storage concepts, to which the Power-to-Gas technology or, more precisely, system belongs. The volumetric storage capacities of the technologies listed in Table 1.1 have been reflected by Bajohr et al. (2011) (Fig. 1.2). Due to the calorific value of methane, a factor 3 higher compared to hydrogen, the volumetric energy storage density of methane is by far the highest of all options depicted in Fig. 1.2. Beside the various possibilities to re-utilize methane, for example as fuel in the mobility sector, or the reconversion to electricity in gas turbine combined cycle plants, the high volumetric density as well as the existing infrastructure for transport and storage are the main advantages of this gaseous, chemical storage media. The main drawback is the efficiency losses of each conversion step. More details follow in Chap. 2.

In general, it has become especially apparent that the current technical and organizational structures of the power supply systems are only partially suitable for the efficient integration of the rapidly growing portion of renewable energies. In the long term, in order to provide a safe and cheap power supply, as well as construction of additional new storage capacities and technologies, an adaptation of the systems is required so that the generation of renewable resources can be coordinated with demand, available grid and storage capacities. This can be achieved with the integration of chemical energy storage such as, for example, Power-to-Gas plants. Through the possible decentralized construction of Power-to-Gas plants alongside the production plants with volatile production patterns, the electrical energy can be saved before input into the grid and in favorable times, transported over the power lines or directly fed into the natural gas network in the form of hydrogen or methane. The new possibilities of energy storage using the Power-to-Gas system will be dealt with in detail in the following chapters which give an introduction to the Power-to-Gas technology and a compact technological description of the central elements of this technology. Subsequently, the economic characteristics as well as the importance of the Power-to-Gas system will be reflected, with the aim to give a brief summary of the state of the art and future challenges for the Power-to-Gas technology.

References

Bajohr S, Götz M, Graf F, Ortloff F (2011) Speicherung von regenerativ erzeugter elektrischer Energie in der Erdgasinfrastruktur. gwf-Gas, Erdgas:200–210

Beaudin M, Zareipour H, Schellenberglabe A, Rosehart W (2010) Energy storage for mitigating the variability of renewable electricity sources: an updated review. Energy Sustain Dev 14:302–314

Chen H et al (2009) Progress in electrical energy storage system: a critical review. Prog Nat Sci 19:291–312

Diaz-Gonzalez F, Sumpe A, Gomis-Bellmunt O, Villafáfila-Robles E (2012) A review of energy storage technologies for wind power applications. Renew Sustain Energy Rev 16:2154–2171

Klaus T et al (2010) Energieziel 2050: 100 % strom aus erneuerbaren Energien. Umweltbundesamt, Dessau-Roßlau

Steinmüller H et al (2014) Power to gas—eine Systemanalyse. Markt- und Technologiescouting und –analyse. Project report for the Austrian Federal Ministry of Science, Research and Economy

Chapter 2
The Power-to-Gas Concept

This chapter gives an overview of the technological fundamentals of the Power-to-Gas concept. After a general introduction to the concept itself, efficiencies and synergy potentials of the Power-to-Gas technology are described. Furthermore, a very short introduction to similar concepts is given, as well as a view to the technological challenges and restrictions for integration of hydrogen into the natural gas grid. Due to the limited available space, only the main aspects are addressed with reference to further reading.

As described in Chap. 1, temporal and spatial fluctuations of the power generation by renewable energy sources demand both high-capacity distribution systems as well as intermittent storage possibilities. The Power-to-Gas concept approaches these demands by the conversion of the electrical power to a gaseous chemical storage medium, the energy-rich gases hydrogen (H_2) and methane (CH_4), respectively. The Power-to-Gas concept is depicted in Fig. 2.1 (Sterner 2009; Grond et al. 2013; Müller-Syring et al. 2013a; Deutsche Energieagentur 2013; Egner et al. 2012).

As shown in the upper left side of Fig. 2.1, renewable electricity is usually transferred to the power grid. The transport of electricity is limited on the one hand by the actual grid-side demand which may result in temporal excess energies. On the other hand, renewable energy production may be located in afield areas with limited transport capacities or completely autarkic structures. According to Fig. 2.1, the renewable electric power is then used in a water electrolysis plant to produce hydrogen and oxygen from water. Oxygen can be released to the atmosphere, or can be preferably used in industrial production processes, like the chemical or the metallurgical industry. But, the utilization of oxygen depends strongly on the local conditions, particularly the distance to the potential consumers and the consumer demand. The actual product is hydrogen which can be transported either in an own hydrogen distribution grid, as admixture in the natural gas grid, by truck or by train. Hydrogen can also be stored in appropriate facilities or together with natural gas in existing natural gas storage infrastructure.

© The Author(s) 2014

M. Lehner et al., *Power-to-Gas: Technology and Business Models*,
SpringerBriefs in Energy, DOI 10.1007/978-3-319-03995-4_2

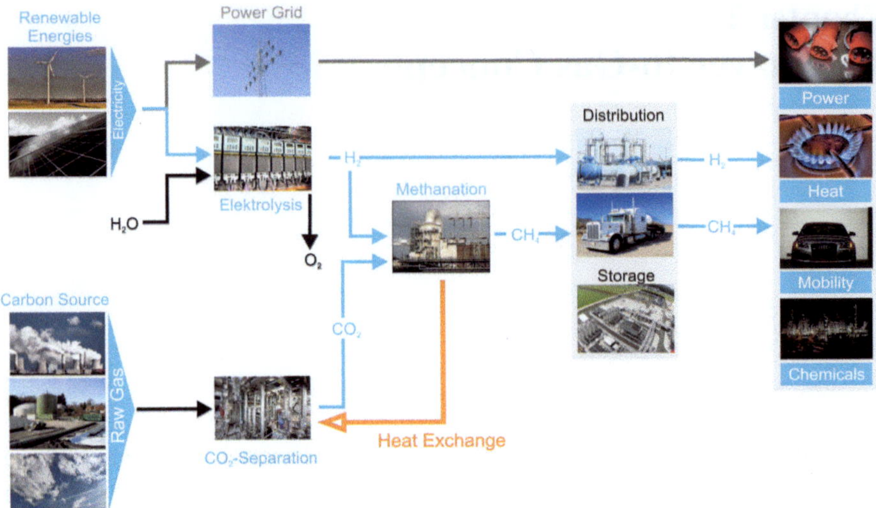

Fig. 2.1 The Power-to-Gas concept

Hydrogen is then re-transferred either to electric power, as fuel in the mobility sector, or as valuable feedstock for industry. Particularly, the chemical, petrochemical and metallurgical industries consume huge annual amounts of hydrogen (approximately 600 billion m^3/a) which are currently produced mainly by methane steam reforming.[1]

Thus, hydrogen is the first possible end-product of the Power-to-Gas process chain. But, the producible volume of hydrogen is limited by either missing hydrogen infrastructure (i.e. hydrogen grid, storage facilities, end-use technologies), or by a maximum allowable content in the natural gas grid.

Therefore, the second, but optional process step within the Power-to-Gas process chain is methanation. Hydrogen and carbon dioxide (CO_2) synthesize to methane, either by a chemically or biologically catalyzed reaction. The produced methane is called synthetic or substitute natural gas (SNG). The by-product of this reaction is steam (H_2O). The necessary carbon dioxide can be derived from exhaust or process gases of industrial production processes or fossil power plants, biogas plants, or in principal also from the atmosphere or from sea water (Fig. 2.1). The latter options are certainly energy-intensive. Since pure carbon dioxide sources are only rarely available (Ausfelder and Bazzanella 2008), carbon capture plays a significant role in the Power-to-gas concept, both technically and economically.

The main advantage of methane as end-product of the Power-to-Gas process chain is its unlimited usability in the gas infrastructure. SNG bi-directionally links the power grid and the gas grid. Already existing transport and storage possibilities

---

[1] The by-product of the reforming process is carbon dioxide. Methane steam reforming is the reverse reaction of methanation, see Chap. 4.

of the gas grid are used for the transfer of renewable electricity in the form of SNG. The huge gas storage facilities in Europe[2] enable the intermittent retention of renewable energy in the range of up to 1,000 TWh. Furthermore, the infrastructure for methane utilization also exists, and is completely technically mature. Beside the conversion into electricity in combined cycle plants, the utilization as fuel in mobility or as feedstock for industry, SNG can be also used for heating. The physical and chemical properties of SNG and natural gas are so similar that no technical changes in the end-use systems have to be undertaken. Almost no new investments in infrastructure for transport, storage and utilization are necessary. This is not only of an economic benefit, but also time saving with regard to permission by authorities, and beneficial for the general public acceptance which is commonly low for any infrastructure projects.

The conversion to the energy-rich gases hydrogen and methane, respectively, enables the transport of the renewable energy outside the power grid, and also the large scale, long-term storage of renewable energy. The chemical energy carriers can be reconverted to electricity, but a multitude of other utilization routes are possible which result in different efficiencies of the total system.

2.1 Efficiencies of Power-to-Gas Process Chains

Since any technical process is associated with energy losses, the high exergy[3] level of electric power is inevitably reduced by the conversion processes within the Power-to-Gas process chain. Therefore, it is preferential to avoid unnecessary conversion steps whenever possible. Electric power should be used as electric power provided that sufficient grid capacities are available. The use of electric power can also be accelerated by generating higher demands, for example by an increased electrification of industrial processes (Leiter et al. 2014). Nevertheless, both the demand management as well as the extension of the power grid are limited, and, therefore, storage of renewable energies is inevitably necessary when the shares of renewable energies are steadily increased in future.

Within the Power-to-Gas process chain, the first usable product is hydrogen. As already mentioned, the chemical, petrochemical and metallurgical industry demands significant volumes of hydrogen. But its utilization requires either a consumer nearby the electrolysis plant, or transport facilities which are poorly developed for hydrogen, at least at present. Furthermore, storage options for hydrogen would enable the buffering and decoupling from the demand side. Beside storage

---

[2] There are about 134 subsurface gas storage facilities throughout Europe with an aggregate storage volume of 94 billion m^3 of natural gas.

[3] Exergy describes the part of energy which is convertible to its full extent in any other form of energy. Anergy is the part of the energy which is not convertible to exergy. The sum of exergy and anergy is the total energy. Electric power consists of 100 % exergy (Baehr 1996).

Table 2.1 Efficiencies for different Power-to-Gas process chains (Sterner et al. 2011)

Path	Efficiency (%)	Boundary conditions
Electricity to gas		
Electricity → Hydrogen	54–72	Including compression to 200 bar (underground storage working pres.)
Electricity → Methane (SNG)	49–64	
Electricity → Hydrogen	57–73	Including compression to 80 bar (feed in gas grid for transportation)
Electricity → Methane (SNG)	50–64	
Electricity → Hydrogen	64–77	Without compression
Electricity → Methane (SNG)	51–65	
Electricity to gas to electricity		
Electricity → Hydrogen → Electricity	34–44	Conversion to electricity: 60 %, compression to 80 bar
Electricity → Methane → Electricity	30–38	
Electricity to gas to combined heat and power (CHP)		
Electricity → Hydrogen → CHP	48–62	40 % electricity and 45 % heat, compression to 80 bar
Electricity → Methane → CHP	43–54	

in underground caverns, the natural gas grid is a potential buffer for hydrogen. The limitations and challenges of the latter option are described later in an own section.

Methanation converts hydrogen to synthetic methane (SNG). The efficiency of the conversion is reported to be 70–85 % in case of the chemical path, greater than 95 % for the biological path (Grond et al. 2013). The main benefit of SNG is its unrestricted compatibility with the natural gas grid, and with the utilization options of natural gas.

The re-powering of methane to electricity in combined cycle plants closes the loop electric power—SNG—electric power. It opens the possibility to produce electric power in areas far away from the renewable power sources, connected by an already existing gas grid. However, the efficiency of this option is the lowest of all possibilities, see Table 2.1.

Slightly better conversion efficiencies can be achieved by producing electricity from hydrogen. Gas turbines, fuel cells or also reverse fuels cells can be utilized for that purpose. Fuel cells would also enable the utilization of hydrogen in the mobility sector, but fuel cell powered cars are technologically not mature, and an infrastructure for hydrogen distribution and storage does not yet exist in most regions.

Generally, the efficiencies for Power-to-Gas systems are increased when the released heat of the systems is used, for example in district heating or in industrial plants nearby (Table 2.1). The pressure level to which the product gases have to be pressurized, has an important influence on the total achievable efficiency. The pressure level mainly depends on the facilities to be used for transport and storage, and is therefore subject to the specific local conditions of a Power-to-Gas plant.

A ranking of the utilization paths according to Table 2.1 cannot be made only by considering the efficiencies alone. Systemic, economic and macroeconomic aspects have to be additionally taken into account, which is the subject of Chap. 5 of this book.

The conversion efficiencies can be improved by either technical progress in the single conversion steps, namely water electrolysis and methanation (see Chaps. 3 and 4 for details), or by synergies with industrial processes which are coupled with the Power-to-Gas plants. Both options are subject of current research activities, see for example (Karlsruher Institut für Technologie 2014; Schöß et al. 2014; Bergins 2014). Some information on synergy potentials is given in the following chapter.

2.2 Plants Sizes and Synergy Potentials

The plant size of a Power-to-Gas system may vary from a few 100 kW connected duty up to several 100 MW, or even in the GW range for autarkic systems. As a consequence, the set-up of the system has to be adjusted individually to the specific boundary conditions of a distinct application. These conditions influence the decision on the end product, hydrogen and methane,[4] respectively, the utilized carbon dioxide source,[5] the use of the potential byproducts, namely oxygen and released process heat, as well as the way the end product is distributed and stored.[6] However, the main purpose of a Power-to-Gas plant may vary: the utilization of (local) excess energies in renewable power production, power grid stabilization or substitution of transport capacities by the natural gas grid, conversion of renewable power for long-term storage, or even the operation of large scale, completely autarkic systems. Furthermore, the desired utilization of the end product influences the plant size and the mode of operation (i.e. pressure level, annual availability etc.). Therefore, future Power-to-Gas systems will consist of completely different plant set-ups, operation modes and plant sizes. Consequently, the Power-to-Gas technology has to be flexible, easily up-scalable and modular in order to allow an adjustment to the specific conditions. Particularly, the total investment and operation costs of a Power-to-Gas plant are influenced by a number of factors, and therefore, the cost for the end product, hydrogen or methane, is not only subject to the actual price of electricity. Indeed, the annual operation hours influence the product costs significantly more than the electricity costs (Kinger 2012). The total cost structure, as well as the technological setup, is also determined by the possibilities to utilize the byproducts (see also Chap. 5).

---

[4] A comprehensive study has been performed recently by DVGW (Müller-Syring et al. 2013b). Four distinct locations for Power-to-Gas plants are examined, and specific Power-to-Gas plant concepts are determined.

[5] Small Power-to-Gas plants (few 100 kW) may use carbon dioxide from biogas plants, and may also utilize biological methanation instead of chemical. For Power-to-Gas plants in the MW scale, industrial carbon dioxide sources are required, and preferably chemical methanation is used.

[6] The way the end product is distributed and stored is mainly a question of the existing infrastructure on site, and the desired utilization of the end product (Müller-Syring et al. 2013b).

In terms of the byproducts of a Power-to-Gas system, the utilization of released reaction heat from the methanation, as well as of oxygen from the electrolysis has to be considered. The methanation reaction is exothermic, and therefore a surplus of heat is generated in the Power-to-Gas process chain. The released heat can be utilized, for example, for the carbon capture process supplying the necessary carbon dioxide for methanation. In case of carbon capture by chemical absorption with amine based solutions, the main energy demand arises from the regeneration of the rich solutions which is done by heating of the scrubbing liquids. An example for the heat integration is given in Chap. 4. It is shown, that the heat surplus of methanation exceeds by far the demand of carbon capture, and therefore can be used additionally for power generation, for example.

The utilization of the byproduct oxygen is coupled to an industrial user, like the chemical or the metallurgical industry. Special power plant concepts, like the oxy-fuel technology, would also need oxygen in significant amounts. Different authors made studies of the implementation of a Power-to-Gas plant in an industrial environment (Schöß et al. 2014; Bergins 2014). It is shown that numerous synergies can be generated. Whereas Bergins (2014) focus on the synergetic coupling with a steel plant, its oxygen demand and the possibility to utilize excess heats, and pronounces the economy of scale for such installations, Schöß et al. (2014) proposes the use of process gases from the steel industry as carbon source. All these concepts are suitable for Power-to-Gas plants on a larger scale (MW). Schöß et al. (2014) simulates an electrolyzer duty of 53.9 MW and a methane production of 4,507 m^3/h.

Figure 2.2 depicts a future vision of a large-scale, autarkic Power-to-Gas system (Frühwirth 2014). An offshore wind park supplies its renewable electricity production

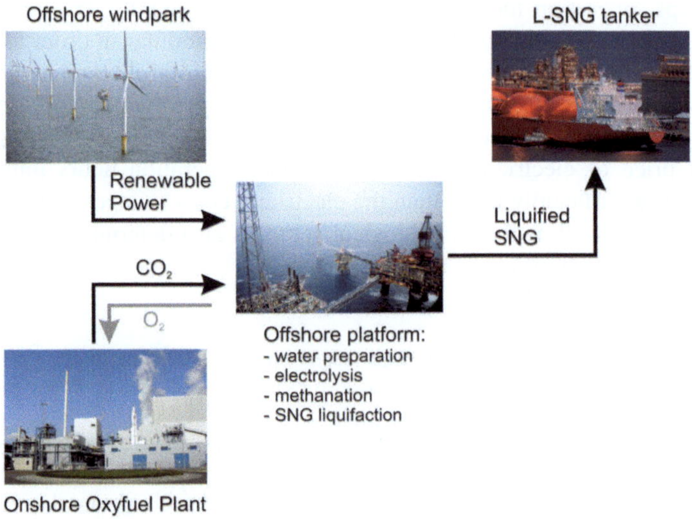

Fig. 2.2 Future vision of an autarkic Power-to-Gas system for the offshore production of liquefied SNG (pictures by courtesy of Statoil, *source* http://fotoweb.statoil.com/fotoweb/Default.fwx)

to an offshore platform nearby. At this platform, an appropriate electrolyzer unit converts the electric power to hydrogen, which is subsequently synthesized to methane. The water for electrolysis is derived from the methanation as well as from evaporated sea water. The necessary evaporation heat is covered by the released reaction heat of methanation. The carbon dioxide is produced in an onshore oxyfuel plant, and pumped through a pipeline to the platform. In return, oxygen from the electrolysis is transferred to the oxyfuel plant. For this reason, the air separation unit at the oxyfuel plant can be saved. Furthermore, the oxyfuel plant can be operated continuously, and it provides the necessary base load in the power grid. Finally, the produced SNG is liquefied to Liquefied Synthetic Natural Gas (L-SNG) at the platform, and is shipped by LNG tankers.

This concept is a kind of "big picture" for the Power-to-Gas technology. It is, of course, far from realization, but it demonstrates future possibilities of a gas "exploration" which is completely on a renewable basis. At the moment a feasibility study of this concept is been worked on (Frühwirth 2014) which should show how mass and energy balances of all parts of such a system fit together. If this concept seems to be feasible, a lot of synergy potential exists.

2.3 Similar Concepts

Power-to-Gas is not the only possibility of converting renewable energy to chemical storage media. Beside the gases hydrogen and methane, other chemical energy carriers, like methanol, formic acid or fuels can be produced. These utilization routes are summarized under the term "Power-to-Liquids", or sometimes also "Power-to-Fuels" (Bilfinger Industrial Technologies 2014). The hydrogen derived from water electrolysis is catalytically converted with CO_2 to methanol, or via Fischer-Tropsch-synthesis to fuels. The basic technological setup is the same as for methanation used in the Power-to-Gas concept. Power-to-Liquids supplies liquid energy carriers which can be easily transported by tankers (road, train, ship), and which are not depending on an appropriate gas grid. As with the products of a Power-to-Gas system, methanol and fuels have an already established utilization, for example in the mobility sector or in the chemical industry. It is discussed that Power-to-Liquid systems are more suitable for larger scale plants whereas Power-to-Gas systems can cover also smaller scales (Leiter et al. 2014). Since Fischer-Tropsch synthesis is not very selective, some additional efforts for the refining of the products result.

Another, very interesting and quite promising concept is introduced under the term "Power-to-Chemistry®". The term is a registered trademark of Evonik Industries (Markowz 2014). Other than in the Power-to-Gas or Power-to-Liquid concept, renewable electricity is not converted to hydrogen in an electrolyzer. An arc furnace is used to convert methane to acetylene and hydrogen according to following fundamental, simplified equation:

$$2CH_4 \rightarrow C_2H_2 + 3H_2 \quad \Delta H_R^0 = +376.8\,kJ/mol \tag{2.1}$$

A byproduct of this reaction is ethylene (C_2H_4). A single arc furnace has a power consumption of 10 MW_{el}, and can be operated highly dynamic with start-up times lower than 1 min. The conversion efficiency is also quite high, 1 MWh_{el} is converted to 0.9 MWh_{th} (Markowz 2013). Load flexibility can be achieved by the parallel connection of a number of arc furnaces. Acetylene was an important intermediate in the chemical industry until the late 1960s, and is nowadays substituted by ethylene and propylene produced in steam crackers. All process routes for the utilization of acetylene are known and still available. Another benefit is the conversion of a C1 hydrocarbon (CH_4) to a C2 hydrocarbon, the "byproduct" hydrogen is also a valuable base chemical. Finally, no carbon dioxide source is required, and thus also the carbon capture costs are saved. But, the concept does not disburden the electricity grid, since the electric power has to be transferred from the renewable source to the chemical site. Additionally, the renewable power is mainly converted to a chemical intermediate, and thus the multiple utilization routes of the Power-to-Gas concept (Fig. 2.1) are not available.

2.4 Integration in the Natural Gas Grid

The products of the chemical conversion in a Power-to-Gas plant, hydrogen and SNG (methane), respectively, have to be preferentially[7] transported by the natural gas grid, and stored in the grid as well as the connected large-scale storage facilities. Therefore, the impacts of the injection of hydrogen or SNG have to be evaluated. Furthermore, the requirements for the injected gas composition and gas volume, as well as any restrictions for the product gas injection have to be considered.

The case of SNG as end product of the Power-to-Gas process chain is less critical than hydrogen, because natural gas consists to a large extent of methane.[8] Therefore, a practically unlimited injection of SNG into the gas grid is possible. Since methanation is an equilibrium reaction, parts of the educt gases, hydrogen and carbon dioxide, are not converted to methane. Furthermore, the product gas mixture emerging from the methanation reactor contains significant amounts of steam, the main byproduct of the methanation reaction. Therefore, a product gas upgrade is necessary in order to meet the requirements for the injection of the produced SNG into the gas grid. Product gas upgrade processes and the specifications for the SNG composition are given in Sect. 4.2 and Table 4.7.

The injection of hydrogen into the natural gas grid raises a number of questions which have been investigated in some recent studies (Müller-Syring et al.

---

[7] A gas grid for hydrogen only exists in a few, spatially limited regions. Therefore, hydrogen should be injected to the existing and well established natural gas grid.

[8] Natural gas qualities are categorized in H- and L-gas (see also Table 2.2). H-gas contains >96 vol.% CH_4, L-gas >88 vol.% CH_4 (Müller-Syring et al. 2013b).

Table 2.2 Specification of gas properties according to different regulations [extended from (Müller-Syring et al. 2013b)]

Parameter	Unit	DVGW G 260 (Weißdruck May 2008)	ÖVGW G31	EASEE-gas	DIN 51624
Wobbe index	kWh/m^3		13.3–15.7		
L—gas		10.5–13.0		–	–
H—gas		12.8–15.7		13.6–15.8	–
Heating value	kWh/m^3	8.4–13.1	10.7–12.8	–	–
Relative density	–	0.55–0.75	0.55–0.66	0.555–0.75	0.555–0.7
Methane number	–	DIN 51624	–	–	70
Hydrogen content	vol.%	≤5	≤4	–	2

2012, 2013b; Melaina et al. 2013; Florisson 2010; Müller-Syring and Henel 2014; Haeseldonckx and D'haeseleer 2007). In detail, following problems have to be considered:

- The influence on the gas characteristics, like Wobbe[9]-index and heating value: the basis for this evaluation is the existing regulations for the natural gas grid, and the required properties of the transported gases, see Table 2.2. With increasing amounts of hydrogen, both Wobbe-index and heating values are reduced. The tolerable percentage of hydrogen strongly depends on the properties of the natural gas quality in the grid. An admixture of 5 % up to 15 % of hydrogen is possible (Müller-Syring et al. 2013b).
- The impacts on the gas infrastructure: piping, controls, fittings, valves, gaskets and the metering systems. Both, steel and plastic piping materials are usually capable of handling admixtures of hydrogen up to 30 % and more. Leakage rates will increase, but are still economically and ecologically tolerable (Müller-Syring et al. 2013b; Florisson 2010). Particularly, the metering systems have to be adjusted for hydrogen admixtures.
- The transport capacities: the volumetric heating value of hydrogen is three times lower than that of methane. Therefore, with the same volume flow of hydrogen, three times lower energy transport is achieved. An admixture of 10 vol.% hydrogen results in a 5–6 % decrease of the transport capacity (Müller-Syring et al. 2013b). But the full transport capacities of gas pipelines are exploited only for a few days per year. Nevertheless, for the transportation of the same energy amounts, a higher volume has to be transported which results in higher pressure losses, and consequently in increased compressor powers. Also, the capability of the installed compressors for the transportation of hydrogen/methane mixtures has to be evaluated.

---

[9] The Wobbe-index is the ratio of the heating value and the square root of the relative density of the gas. The relative density is the ratio of the gas density to the density of air under standard conditions. The burner power remains constant with same Wobbe-indices despite different heating values (Müller-Syring et al. 2013b).

- The impacts on end user infrastructure: domestic appliances, like heating systems for houses or apartments, can usually operate hydrogen admixtures of up to 20 %, eventually the adaption of the burner nozzles is necessary due to the higher flame velocities (Müller-Syring et al. 2012).
  Gas turbines are more sensitive to hydrogen. Most of the manufacturers limit the hydrogen content to 1 or 2 vol.%, but laboratory tests show the possibility of admixtures up to 14 % (Müller-Syring et al. 2012). Similar considerations are valid for gas motors.
- The impacts in the automotive sector: the methane number (see also Table 2.2) is reduced by the admixture of hydrogen, 10 % hydrogen results in a decrease of 5–7 units. An exceedance of the knocking limit may be the consequence. But, the limitation of the hydrogen content to 2 vol.% of DIN 51624 is much more critical. The background of this limit is the lack of knowledge how the steel storage tanks, both in the cars and the filling stations, can tolerate higher hydrogen contents in a long-term view (Müller-Syring et al. 2013b).
- The impacts on underground gas storage facilities: for the storage of natural gas, salt caverns and depleted gas reservoirs are currently operated. Particularly, for porous subsurface reservoirs some fundamental questions are still open, for example microbiological reactions in the reservoir, de-mixing processes or the general impacts on the geochemical conditions. Those questions are addressed in a research project, currently carried out by a consortium of industry and universities in Austria.[10]

The maximum percentage of hydrogen is also limited by the natural gas flow at the distinct location where it is injected into the grid. Those grid sections with low annual turn-over of gas are less suitable for the introduction of hydrogen (Müller-Syring et al. 2013b).

Currently, it is recommended to limit the concentration of hydrogen in the natural gas grid to 2 vol.% in case natural gas filling stations are connected to the gas grid, and 10 vol.% H_2, in case no natural gas filling station, no gas turbine or gas motor are connected to the natural gas grid (Müller-Syring and Henel 2014).

References

Ausfelder F, Bazzanella A (2008) Verwertung und Speicherung von CO_2. Dechema, Frankfurt/ Main

Baehr HD (1996) Thermodynamik, 9th edn. Springer, Berlin, p 134

Bergins C (2014) Energiewende umsetzen mit dem Großanlagenbau. Presentation at ProcessNet-Fachgruppe "Energieverfahrenstechnik", Karlsruhe, Hitachi Power Europe, 18, February 2014

Bilfinger Industrial Technologies (2014) Power-to-liquids. http://www.sunfire.de/wp-content/uploads/BILit_FactSheet_POWER-TO-LIQUIDS_EMS_en.pdf. Accessed 25 May 2014

---

[10] http://www.underground-sun-storage.at/en.html. Accessed 22 May 2014.

Deutsche Energieagentur (2013) Strategieplattform Power-to-Gas. Thesenpapier: Technik und Technologieentwicklung. http://www.powertogas.info/fileadmin/user_upload/downloads/Positionen_Thesen/PowertoGas_Thesenpapier_Technik.pdf. Accessed 7 May 2014

Egner S, Krätschmer W, Faulstich M (2012) Perspektiven der Energiewende. In: Lorber K et al. (eds) DepoTech 2012, Tagungsband zur 11. DepoTech Konferenz, Leoben, pp 49–56

Florisson O (2010) Naturally preparing for the hydrogen economy by using the existing natural gas system as catalyst. Final publishable activity report, N.V. Nederlandse Gasunie

Frühwirth V (2014) Feasibility study of a large scale power-to-gas system. Master thesis, Montanuniversität Leoben (in preparation)

Grond L, Schulze P, Holstein S (2013) Systems analyses power to gas: deliverable 1: technology review. DNV KEMA energy & sustainability, Groningen

Haeseldonckx D, D'haeseleer W (2007) The use of the natural-gas pipeline infrastructure for hydrogen transport in a changing market structure. EHEC2005 32(10–11):1381–1386

Karlsruher Institut für Technologie (ed) (2014) Power-to-Gas: Wind und Sonne in Erdgas speichern. http://www.kit.edu/downloads/pi/KIT_PI_2014_044_Power-to-Gas_-_Wind_und_Sonne_in_Erdgas_speichern_.pdf. Accessed 23 May 2014

Kinger G (2012) Green energy conversion and storage (Geco). Endbericht for FFG project 829943, Wien

Leiter W, Schüth F, Wagemann K (2014) Diskussionspapier: Überschussstrom nutzbar machen. Dechema, Frankfurt. http://dechema.wordpress.com/2014/01/31/ueberschussstrom_1/. Accessed 15 May 2014

Markowz G (2013) Power-to-Chemistry® Ein alternatives Konzept zur chemischen Energiespeicherung. Presentation at Dechema Kolloquium „Wind-to-Gas" Frankfurt, Evonik Industries, 7 March 2013

Markowz G (2014) Power-to-Chemistry® Strom speichern im industriellen Maßstab. Presentation at Innovationskongress Berlin, EVONIK Industries, 7 May 2014

Melaina MW, Antonia O, Penev M (2013) Blending hydrogen into natural gas pipeline networks: a review of key issues. National Renewable Energy Laboratory, Golden

Müller-Syring G, Hüttenrauch J, Zöllner S (2012) Erarbeitung von Basisinformationen zur Positionierung des Energieträgers Erdgas im zukünftigen Energiemix in Österreich: AP 2: Evaluierung der existierenden Infrastrukturen auf Grundlage der ermittelten Potentiale. Abschlussbericht, Leipzig

Müller-Syring G et al (2013a) Entwicklung von modularen Konzepten zur Erzeugung, Speicherung und Einspeisung von Wasserstoff und Methan ins Erdgasnetz, DVGW Bericht zu Fördezeichen G 1-07-10:119

Müller-Syring G et al (2013b) Entwicklung von modularen Konzepten zur Erzeugung, Speicherung und Einspeisung von Wasserstoff und Methan ins Erdgasnetz. Abschlussbericht, DVGW Förderkennzeichen G1-07-10, Bonn

Müller-Syring G, Henel M (2014) Auswirkungen von Wasserstoff im Erdgas in Gasverteilnetzen und bei Endverbrauchern. Gwf Gas-Erdgas:310-312

Schöß MA, Redenius A, Turek T, Güttel R (2014) Chemische Speicherung regenerativer elektrischer Energie durch Methanisierung von Prozessgasen aus der Stahlindustrie. Chem Ing Tech 86(5):734–739

Sterner M (2009) Bioenergy and renewable power methane in integrated 100 % renewable energy systems. Dissertation, Universität Kassel

Sterner M, Jentsch M, Holzhammer U (2011) Energiewirtschaftliche und ökologische Bewertung eines Windgas-Angebotes. Fraunhofer IWES, Kassel, p 18

Chapter 3
Water Electrolysis

Abstract This chapter reviews basic aspects of water electrolysis technologies. First, fundamentals of water electrolysis are discussed to give an overview on basic modes of operation, different ways to determine the electrolyzer efficiency and basic aspects of performance optimization strategies. Second the three main water electrolysis technologies, namely the alkaline electrolysis (AEC), the polymer electrolyte membrane electrolysis (PEMEC) and the solid oxide electrolyte electrolysis (SOEC), are described in more detail. The state of the art, typical system setups, operating characteristics, main component materials, technological assets and drawbacks, current and future developments and future challenges of each of the main technologies are discussed.

3.1 Introduction

The Greek expression "Lysis" means decomposition and analogous electrolysis describes a decomposition process in which electrical energy is the main driving force for participating chemical reactions. In the case of water electrolysis a voltage and a direct current are applied to water, what causes dissociation of water molecules into the product gases hydrogen and oxygen. Therefore a water electrolyzer is basically an electrochemical device that converts electrical (in some cases also thermal) energy into chemical energy. The storage medium hydrogen is currently of main economic interest.

Hydrogen, the simplest and lightest element in the periodic table, is a colorless, odorless, tasteless and nontoxic gas. Hydrogen, with a worldwide annual production of about 55 million metric tons, is primarily an industrial resource used e.g. in ammonia and methanol production, refineries, chemical, electronic, metal, glass and food industry. Around 95 % of all hydrogen is obtained from fossil fuels but just about 4 % are provided by electrolysis to date (Holladay et al. 2009).

Hydrogen has a very high energy density by weight (33.3 kWh/kg), which is up to three times larger compared to liquid hydrocarbon based energy carriers. Hydrogen is not an energy source but it is a secondary energy carrier that offers a

© The Author(s) 2014

M. Lehner et al., *Power-to-Gas: Technology and Business Models*,
SpringerBriefs in Energy, DOI 10.1007/978-3-319-03995-4_3

wide range of benefits, which are receiving great attention nowadays. The growing interest is mainly driven by the facts that hydrogen offers maybe the greatest long-term potential as an alternative fuel, as an all-purpose energy carrier and as an energy storage medium.

Our energy system is currently changing significantly due to the continuing, massive integration of renewable energies. This transformation process and thereby especially the intermittency of wind and solar power production gives rise to new challenges like e.g. system operation, load leveling, distributed generation management or storing and utilization of surplus energy. Hydrogen produced by water electrolysis can help to cope with these challenges as it offers the chance to store and transport electrical energy that can be used in various economic sectors independent of time. This approach is currently discussed under the synonym Power-to-Gas.

Water electrolysis plays a central role in Power-to-Gas systems as it represents the linkage between electrical and chemical energy, independent if the produced hydrogen is used in its elemental form or as an intermediate for further chemical reactions. The most important demands on electrolyzers for Power-to-Gas systems are highly dynamic modes of operation, wide partial load ranges with sufficiently high efficiencies and satisfying gas purity levels, compact stack designs, high unit power densities, high production capacities and low investment respectively operating costs. Although water electrolysis is already a well-established technology, further improvements are required to meet those requirements. Currently a lot of fundamental and applied research and development efforts are carried out to pave the way for a broader implementation of electrolytic hydrogen production into the market and to facilitate a larger integration of the Power-to-Gas technology into the electrical grid.

3.2 Historical Background

Water electrolysis is a rather old process known for about 200 years. It is not quite clear who first discovered water electrolysis since quite different statements can be found in the literature. However, Trasatti (1999) and De Levie (1999) pointed out that first the two Dutchmen Adriaan Paets van Troostwijk (1752–1837) and Jan Rudolph Deiman (1743–1808) observed the decomposition of water into a mixture of "combustible air" and "life-giving air" caused by electric discharges back in 1789. In 1800 the two Englishmen William Nicholson (1753–1815) and Anthony Carlisle (1768–1840) observed the phenomena of water decomposition initiated by direct current. The basic physical law of electrolysis was discovered about 30 years later (1834) by the English scientist Michael Faraday (1791–1867).

This was the starting point for the gradual development of industrial electrolyzers (Kreuter and Hofmann 1998). In 1902 worldwide more than 400 industrial electrolyzers were already in operation. In 1939, the first large electrolysis plant with a capacity of 10,000 m^3 H$_2$/h, built by the Norwegian company Norsk Hydro Electrolyzers, went into operation. In 1948 the first pressurized electrolyzer was manufactured. The first solid polymer electrolyte membrane electrolyzer was built in 1966. A couple of years later (1972) the development of solid oxide electrolyte electrolyzers was started.

3.3 Thermodynamics of Water Electrolysis

The overall equation of the basic water splitting reaction is noted as below:

$$H_2O \rightarrow 1/2\ O_2 + H_2 \tag{3.1}$$

$\Delta H(T)$ is the total amount of energy that has to be supplied to an electrolysis cell in order to split water molecules according to reaction 3.1. The change of Gibbs free energy $\Delta G(T)$ represents the amount of electrical energy and $T\Delta S(T)$ represents the amount of heat, which have to be supplied to an electrolysis cell to drive the water splitting reaction.

$$\Delta H(T) = \Delta G(T) + T\Delta S(T) \tag{3.2}$$

The minimum applied cell potential for starting a water splitting reaction is represented by the reversible voltage V_{rev}, which is related to the Gibbs free energy change as follows:

$$V_{rev} = \frac{\Delta G}{nF} = 1.23\ \text{V} \tag{3.3}$$

With ΔG of 237.22 kJ/mol (at standard conditions of 1 bar and at 298 K), n the number of electrons transferred in reaction 3.1 and the Faraday constant F of 96,487 C/mol the reversible voltage calculates as 1.23 V.

The thermo-neutral voltage V_{th}, defined by Eq. 3.4, is related to the enthalpy change associated with the water splitting reaction.

$$V_{th} = \frac{\Delta H}{nF} = \frac{\Delta G}{nF} + \frac{T\Delta S}{nF} = 1.48\ \text{V} \tag{3.4}$$

With ΔH of 285,84 kJ/mol and a temperature T of 298 K a thermo-neutral voltage of 1.48 V is obtained.

If the voltage E_{cell}, which is applied to the electrolysis cell, is higher than V_{rev} but lower than V_{th} water splitting just takes place by absorbing heat from the environment as the cell dissipates the heat associated with the change in entropy irreversibly. If $E_{cell} = V_{th}$ the Joule heat generated within the electrolysis cell equals the heat consumption of the endothermic electrolysis reaction and therefore no heat exchange with the environment would be required. If $E_{cell} > V_{th}$ the electrolysis cell produces surplus heat due to joule heating and has to be properly cooled in order to reduce degradation of the system.

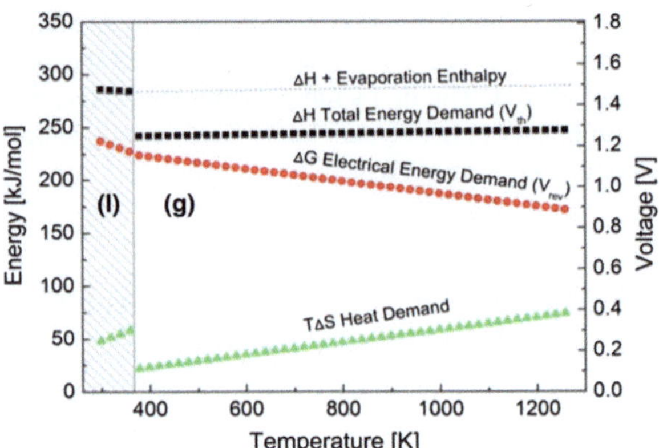

Fig. 3.1 Water electrolyzer thermodynamics and cell-voltage as a function of operation temperature. [Data for calculation taken from Dorf (2004)]

The operating temperature and pressure are important parameters for electrolyzer systems and have to be carefully chosen. Figure 3.1 shows the temperature dependence of the particular energy demands and the corresponding cell voltages for water electrolysis from a pure thermodynamic point of view. Due to vaporization of water at 373 K the total energy demand (ΔH resp. V_{th}) curve is discontinuous and remains almost unchanged within a particular state of matter. The figure clearly shows that if steam is available as feedstock, steam electrolysis requires less energy compared to liquid water electrolysis. The electrical energy demand (ΔG resp. V_{rev}) continuously decreases with increasing temperature resulting in a ~30 % reduction by switching from 273 to 1,273 K.

The influence of pressure on the cell voltage is small and can be estimated with a rewritten form of the well-known Nernst Equation as follows:

$$\Delta V = V - V^0 = -\frac{RT}{nF} \ln \frac{1}{\sqrt{P}} \tag{3.5}$$

R is the ideal gas constant (8.314 J/mol K) and P the overall pressure within the electrolysis cell assumed to be equal at both electrodes. An increase in the overall pressure from 1 to 200 bar corresponds to an increase of the theoretical cell voltage V by just 34 mV at 298 K and by 122 mV at 1,073 K respectively. Although raising the operating pressure causes an increase of the theoretical reversible voltage by a few percent it has various positive, system-relevant effects on e.g. the operating voltage, current densities as well as on the production costs of compressed hydrogen.

3.4 Electrolyzer Efficiency

There are quite different definitions of water electrolysis efficiencies available. There are e.g. energy efficiencies, voltage efficiencies, current efficiencies respectively efficiencies based on cell-level, stack-level or system-level. For comparing different systems or technologies it is necessary to make sure which types of efficiencies have been used.

The energy conversion efficiency of an electrolyzer system is generally defined by Eq. 3.6 and includes the overall energy demand of all components of an electrolyzer system.

$$\eta_{sys} = \frac{Energy\ Output}{Energy\ Input} \rightarrow \frac{Heating\ Value\ H_2}{Electrical\ Energy\ Input} \tag{3.6}$$

The energy output usually just refers to the heating value of hydrogen, which can be defined by the higher heating value (HHV $= 3.54$ kWh/scm,[1] scm $=$ standard cubic meter) or the lower heating value (LHV $= 3$ kWh/scm) of H_2. Due to the fact that liquid water is usually used as the feedstock, the energy required for evaporation of water has to be taken into account. Therefore the higher heating value of H_2 should be used for calculating the system efficiency. In conventional water electrolysis the energy input is usually limited to electrical energy (except the operating voltage would be lower than V_{th}). In terms of electrical energy input it is important whether it is provided to a single cell, a set of cells called stack or the entire electrolyzer system including all the auxiliary equipment required to run the system. Additionally the particular system/cell utilization impacts the energy conversion efficiency. Reducing the electrolyzer system utilization below 50 % of rated power typically results in an energy efficiency reduction of 10–30 %. This becomes particularly important for intermittent electrical energy inputs, which are typical for Power-to-Gas applications.

At cell or stack level an electrical efficiency η_{el} (Eq. 3.7) can be defined as a product of a voltage efficiency η_V times a current (faradaic) efficiency η_F. η_V is defined as the ratio of V_{th} and the applied voltage V_{app}, while η_F is defined as the ratio of the measured amount of produced hydrogen n_{meas} and the theoretical amount of produced hydrogen n_{th} according to Faraday's Law. Due to the fact that η_F is over a rather wide range of current densities usually close to one, the electrical efficiency η_{el} approximately equals the voltage efficiency η_V.

$$\eta_{el} = \eta_V * \eta_F = \frac{V_{th}}{V_{app}} * \frac{n_{meas}}{n_{th}} \rightarrow \eta_V \tag{3.7}$$

Since η_{el} of conventional electrolyzers is significantly lower than 1, the operating voltages of conventional water electrolyzers exceed 1.48 V by far due to irreversible losses like internal resistances and overvoltages. In addition those parameters change with varying current densities. Therefore the operating voltage depends on the specific setup of the electrolyzer as well as on the particular mode of operation. The operating cell voltage can be expressed as a sum of the reversible voltage

---

[1] [scm] $=$ standard cubic meter—a cubic meter of gas under standard conditions, defined as an atmospheric pressure of 1.01325 bar and a temperature of 15 °C.

(V_{rev}), which has been described before, and several additional overvoltages as follows:

$$V = V_{rev} + V_{act} + V_{ohm} + V_{conc} \tag{3.8}$$

The activation overvoltage V_{act} is attributed to limited electrode kinetics at the anode and at the cathode. The more complex oxidation reaction at the anode dominates V_{act}. The higher the electro-catalytic activities of the particular electrodes respectively catalyst systems are, the lower are the corresponding activation overpotentials. V_{act} shows a logarithmic dependence on the current density and therefore becomes almost constant at higher current densities.

The overvoltage V_{ohm} is caused by ohmic resistances and is mainly proportional to the electric current passing through the cell. Those losses are caused by resistances to the flow of electrons as well as ions through the particular sections of the cell. V_{ohm} shows a linear dependence on the current density according to Ohms Law and therefore it becomes more significant and sometimes even dominant at higher current densities.

The concentration overvoltage V_{con} is caused by mass transport limitations of mainly gaseous products. It can be minimized by an optimal geometric cell design. V_{con} typically represents the lowest overvoltage of all three presented here.

For a given electrolyzer setup and a certain current density, the total overvoltage decreases with increasing temperature mainly due to improved overall kinetics. Raising the operation temperature also has a positive thermodynamic effect on ΔG, hence on the reversible voltage, as described in the previous section. Due to these various reasons the resulting operating voltage can be significantly reduced with increasing temperature. The operating pressure hardly influences thermodynamics or kinetics of the electrolysis process and thus is less relevant at the cell efficiency level. The operating pressure becomes inherently more important looking at the system efficiency level as described later on.

3.5 Alkaline Electrolyzers

3.5.1 Working and Design Principles

Alkaline electrolyzers represent the most developed water electrolysis technology to date. AEC electrolyzers are currently the standard systems for industrial large-scale electrolysis applications. As depicted in Fig. 3.2, an AEC cell is basically composed of two electrodes, which are fully immersed in an 20–40 wt% aqueous potassium hydroxide (KOH) electrolyte with a microporous diaphragm separating the anodic and cathodic regions. The electrodes are usually made of nickel or nickel plated steel. KOH is preferred over sodium hydroxide (NaOH) electrolytes due to its higher conductivity. The electrolysis cell is housed in a compartment, which is usually made of steel. Product gas leaving the cell is separated from remaining electrolyte, which is then pumped back into the cell. The liquid

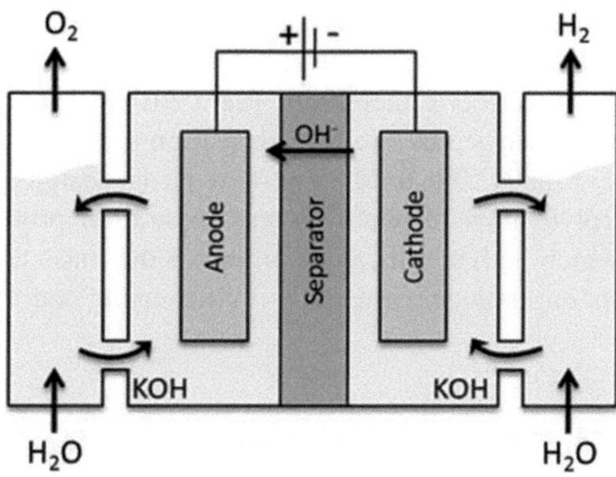

Fig. 3.2 Schematic of the operating principle of an alkaline electrolysis cell

electrolyte is not consumed but has to be replenished over time due to various types of losses.

Applying a direct current to the alkaline electrolyzer cell results in evolution of hydrogen and hydroxide-ions at the cathode according to the half-cell reaction 3.9. The hydroxide ions are migrating through the micro-porous separator and are oxidized at the anode according to the half-cell reaction 3.10. During cell operation, water is consumed but not KOH and therefore water has to be supplied continuously (neglecting physical electrolyte losses).

$$\text{Cathode}\quad 2H_2O + 2e^- \rightarrow H_2 + 2OH^- \tag{3.9}$$

$$\text{Anode}\quad 2OH^- \rightarrow 1/2O_2 + H_2O + 2e^- \tag{3.10}$$

The connection of single electrolysis cells forming a stack, can either be done in parallel (unipolar electrolyzer), or it can be carried out as a serial connection of adjacent single cells (bipolar electrolyzer). Although bipolar electrolyzers are more complex and demand higher manufacturing precision, they are nowadays preferred over unipolar versions due to their significantly lower ohmic losses. Conventional electrolyzers are commonly composed of 30–200 single cells with an effective membrane area of each cell in a range of up to 1–3 m^2.

Another important design issue concerns the gap between the electrodes and the separator. The smaller the gap, the lower the ohmic cell resistance caused by limited electrolyte conductivity and evolving gas bubbles. The size of these gaps can approach zero. So called zero-gap systems are currently under development and aim at eliminating bubbles from the critical intra-electrode zone. Drawbacks of this configuration are increased probability for inducing sparks and higher manufacturing standards. Zero-gap configurations can be achieved by inserting electrolyte absorbing layers into the intra-electrode zone or by using gas diffusive electrodes that directly touch the separator (Marini et al. 2012). Another attractive zero-gap approach, which requires a radical redesign of AEC systems, is the use of anion exchange membranes instead of liquid electrolytes and conventional separators (Pletcher and Li 2011). Gas diffusive electrode structures are mechanically pressed against the anion exchange membrane or fabricated directly onto the membrane surface. This technique is commonly known from typical polymer electrolyte membrane fuel cells (PEMFC) and electrolyzers (PEMEC), which are described in a following section in more detail.

Finally, the third important basic design factor is related to an optimal electrolyte flow respectively gas separation procedure in liquid electrolyte systems, which both significantly influence the mass transport properties inside the system. Its importance increases with increasing operating current densities.

3.5.2 Operating Conditions, Performance and Capacities

Conventional AEC systems are usually operated at current densities in the range of 300–500 mA/cm^2 and at corresponding cell voltages in the range of 1.9–2.4 V. The operating temperatures are commonly in the range of 70–90 °C. A predominant portion of installed alkaline electrolyzers are working at atmospheric pressure. Pressurized systems are usually operated at up to 15 bars but seldom above that level. The production capacity of commercially available electrolysis systems covers a wide range of 1–760 scm H_2/h. The largest facilities, comprising several single systems, show total capacities of 10,000+ scm H_2/h. The hydrogen purity is generally at least 99.5+ %.

The system efficiencies greatly vary with system size and also depend e.g. on the particular purity and pressure levels. Typical system efficiencies based on the HHV of H_2 are in the range of 60–80 % corresponding to specific energy demands of 4.3–5.5 kWh/scm H_2. Electrolyzers operated at atmospheric pressure are slightly more efficient compared to pressurized ones. This becomes gradually less important with increasing system size.

In terms of dynamic operation the conventional alkaline electrolyzers can be typically operated at ~20–100 % of rated power, while operation in the lower half of that range usually results in significantly reduced gas quality and increasingly reduced system efficiencies. Conventional systems tend to have long startup times (minutes to hours, depending whether from stand-by or cold-start) and usually they have difficulties to follow rapidly changing power inputs.

3.5.3 Cell Components

Basic containment materials, separator plates and current distributers are typically made of Ni, Ni-plated steel or Ni-plated stainless steel. A proper sealing is guaranteed by polymeric or metallic materials. This is pretty much standard and current research activities mainly focus on the development of new separator membranes, highly active and durable electrodes respectively solid electrolytes.

The separator membrane has to guarantee a sufficient separation of the electrolyte and the product-gas between the oppositely charged electrode regions by increasing the cell resistance as less as possible. Furthermore a selective ion transport has to be guaranteed. In the past asbestos was almost exclusively used as separator material. However, due to problems with corrosion at elevated temperatures and due to its severely adverse health effects many different alternatives to asbestos have been developed over the past decades. Current diaphragms are mainly based on sulfonated polymers, polyphenylene sulfides, polybenzimides and composite materials thereof (Otero et al. 2014). Especially Zirfon, which consists of 60–80 wt% ZrO_2 in a polysulfone matrix, is widely studied (Vermeiren 1998) and of particular commercial interest. Further composite materials containing e.g. TiO_2 or Sb_2O_5 particles embedded in various types of polymers have shown good performance as well (Modica et al. 1986).

The electrodes have to show a high catalytic activity and they have to be as durable as possible. High electro-catalytic activity of an electrode is mainly achieved by an appropriate choice of materials and a proper surface nanostructure in order to achieve large electro-active surfaces. The surface is often further activated by additional procedures. Those surface modification and activation processes can be done ex situ or in situ by various methods like etching, sintering, sandblasting, composite coatings, spray coatings or electrodeposition. To facilitate the escape of gas bubbles from the reaction zones usually perforated electrodes with a perforation diameter in the range of 0.1–1 mm are used. Generally electrodes consist of a carrier material, which can be either bulk type or in form of grids or foams.

The most common electrode material is Ni because it represents currently the best compromise between stability, favorable activity and comparably low costs. Thereby Raney-Ni, which is a Ni–Al alloy, is often used as a starting material where superficial Al is leached out in alkaline solution giving a porous surface morphology. However, the deactivation of Ni over time still remains a serious problem. Therefore stabilizing coatings or alternative electrode materials are currently under investigation. Various promising alloys for cathodes like e.g. Ni–Mo, $Ni–MoO_x$, Ni–Fe, Ni–Co, Ni–V, Ni–S, Fe–Co (Subbaraman et al. 2012; Zeng and Zhang 2010; Pletcher et al. 2012; Kaninski et al. 2009) and for anodes like e.g. Co_3O_4, $NiCo_2O_4$, $LaNiO_3$, La–Ca–CoO_3, La–Sr–Co_3O_4 (Subbaraman et al. 2012; Lal et al. 2005; Suntivich 2011; Singh et al. 2007) have been investigated recently.

3.5.4 Technology Status and Challenges

Alkaline water electrolysis is a mature technology that is currently standard for industrial, large-scale, electrolytic hydrogen production at a MW-scale. The key advantages of this technology are its proven durability, maturity, availability, no PGM containing component materials and the comparatively low specific costs.

Two critical key disadvantages of alkaline electrolyzers are the low current densities and the low operating pressures. The current density significantly influences the specific system size and the hydrogen production costs. Therefore it is of particular importance. Improved catalytic activities of the electrodes, advanced electrode designs and optimized separators as well as raising the system pressure are topics of current R&D activities, aiming at increasing the current densities by a factor of 1.5–2. For many applications, especially when the produced hydrogen has to be stored or transported, external compressors are required to compress the produced hydrogen. This adds additional costs and complexity to appropriate systems. Therefore the advantage of raising the operating pressures is manifold. An increase up to 60 bar is a general goal of current developments.

With respect to system durability, typical degradation rates of 1–3 $\mu V/h$ are offering tens of thousands of hours of operation and a regular general overhaul every ~10 years. This satisfies industrial requirements already quite well. The currently typical system efficiencies, especially of big systems, are also at a fairly high level as already mentioned before.

All this holds for conventional, industrial applications under widely constant operating conditions respectively rather constant H_2-production levels. In the course of Power-to-Gas applications, electrolyzers are coupled to renewables that predominantly supply intermittent power. So far this dynamic operation commonly results in lower gas quality, lower system efficiency, more frequent system shut-downs and generally reduced durability of the system. The following of quick load variations is not limited by the kinetics of participating electrochemical reactions but the inertia of auxiliary system components. Recent reports showed that advanced alkaline systems, which are specially designed for intermittent power applications, are able to provide an extended dynamic range of ~10–100 % of rated capacity and improved response times in the lower seconds-range. Relatively long cold start times, the necessity of holding currents during stand-by and gas purity problems during partial load periods are still some of the most critical issues for intermittent operation of alkaline electrolyzers. However, the implications on lifetime of such intermittent operation remain widely unknown and elucidation of those complex problems is subject of various current research projects. In addition to that, these advanced systems are only available on a small scale and have to be up scaled as other electrolysis technologies have to as well.

The specific investment costs for alkaline systems in $€/kW_{el}$ predominantly depends on the system size and the operating pressure. Pressurized systems are roughly estimated 20–30 % more expensive compared to atmospheric systems over a wide range of system sizes. Raising the capacity of electrolysis systems from the kW_{el} to $MW_{el}+$ range, results in a reduction of investment costs by a factor of ~2.5–3. This allows a rough estimation of specific investment costs of around 1,000–1,300 $€/kW_{el}$ on average. The electrolysis stack accounts generally for 50–60 % of the total system costs. This is true for basic system configurations. Upgrading the system by components like e.g. enhanced purification systems, compressors, more efficient AC/DC converters, etc. can easily add additional 25–50 % to the basic costs. For alkaline technology it is generally estimated that cost reductions in the future will be mainly driven by economies of scale rather than by the further developments of particular components.

In summary alkaline electrolyzers are based on a technology that is highly developed, scaled up, proven and comparatively cheap. Low current densities and limited modes of dynamic operation are currently major limitations of that technology. To make this technology fully compatible to Power-to-Gas applications further research has to be carried out.

3.6 Polymer Electrolyte Membrane Electrolysis

3.6.1 Working and Design Principles

PEMEC electrolyzers represent the second important water electrolysis technology. It is generally less developed compared to AEC systems and up to now it is commercially mainly used for small scale niche-applications. Due to the growing interest in water electrolysis systems in general and the opportunity to overcome some severe restrictions of conventional AEC technology, the interest in the PEMEC technology is currently gaining a lot of attention.

A schematic diagram of a PEMEC cell is depicted in Fig. 3.3. In PEMEC cells, thin (~50–250 μm) proton conducting membranes are used as solid polymer electrolytes rather than liquid electrolytes typically used in conventional AEC electrolyzers as described in the previous section. An assembly of such a membrane and an electro-catalytic layer at each side of it is commonly referred to as a membrane electrode assembly (MEA). The MEA represents the core element of a PEMEC cell and is electrically connected via porous current collector layers to cell-separator plates, which often contain flow field patterns for optimal mass transport. Pure water (18 MΩ cm), which is usually fed into the anode compartment, travels along a patterned bipolar plate (separator plate) and/or diffuses through the current collector respectively gas diffusion layer towards the catalytic zone, where the oxidation reaction according to chemical Eq. 3.12 takes place. The hydrogen ions are transported across the proton exchange membrane towards the cathode side, where hydrogen is generated according to reaction Eq. 3.11.

$$\text{Cathode} \quad 2H^+ + 2e^- \rightarrow H_2 \tag{3.11}$$

$$\text{Anode} \quad H_2O \rightarrow 1/2\ O_2 + 2H^+ + 2e^- \tag{3.12}$$

The connection of single cells, forming a stack, is exclusively done in series (bipolar electrolyzer) by filter press construction method. Commercially available electrolyzer stacks are commonly composed of up to 60 single cells with a typical effective membrane area of each single cell in range of 100–300 cm^2, which is at least about a factor of 5–10 smaller compared to AEC systems.

Due to the lack of a liquid electrolyte and all the associated equipment (pumps, gas separation etc.) a solid electrolyte electrolyzer generally allows a significantly more compact system design.

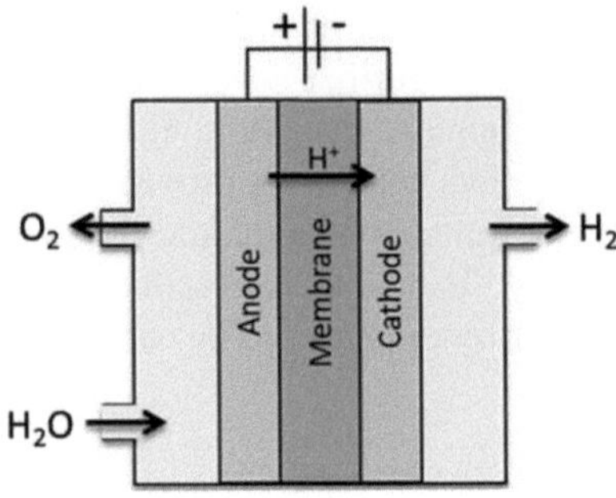

Fig. 3.3 Schematic of the operating principle of a PEMEC electrolysis cell

3.6.2 Operating Conditions, Performance and Capacities

PEMEC systems are usually operated at current densities of 1–2 A/cm^2, which is about a factor of 4 higher compared to AEC technology. The corresponding voltages are in the range of 1.6–2 V. In the lab current densities as high as 5–10 A/cm^2 at cell voltages less than 2.5 V have already been shown. The system efficiencies based on the HHV of H$_2$ are typically in the range of 60–70 %. The operating temperatures are mainly in the range of 60–80 °C. Most of PEMEC systems are working at elevated pressure levels of 30–60 bar, without additional compression units. A few systems even deliver H$_2$-pressures of 100–200 bar without the use of external compressors.

The production capacities of currently commercially available PEMEC systems are typically in the range of 1–40 scm H$_2$/h. The hydrogen purity levels are at least 99.99+ %, where oxygen coming from the anode side is the main impurity.

PEMEC systems can be operated in a highly dynamic fashion covering almost the whole range of 0–100 % of rated power and being able to follow power fluctuations within 100s ms.

3.6.3 Cell Components

A common problem of PEMEC systems is the high acidity of the electrolyte membrane, which is roughly comparable to a 1 M sulfuric acid solution. Furthermore, the high applied voltages at high current densities limit the choice of cell component materials to scarce and expensive materials. Such harsh conditions are in general a challenge for the development of new stack-materials.

The most commonly employed membrane for PEMEC is Nafion, a proton conducting membrane based on perfluorosulfonic acid (Ito et al. 2011). Nafion shows quite good mechanical and electrochemical stabilities, low gas crossover rates and high proton conductivities in the range of around 0.1 S/cm. Major drawbacks are its high costs and its water-assisted proton-conduction mechanism, which limits the operation temperature to <80 °C. In order to overcome these limitations alternative polymers like sulfonated poly(ether ether ketones) (PEEK), poly(ether-sulfones) (PES) and sulfonated polyphenyl quinoxaline (SPPQ) are currently under investigation (Mittelsteadt et al. 2012). So far alternative polymer electrolytes generally suffer from lower apply able current densities and lower durability's compared to Nafion. However, even Nafion does not fully satisfy stability requirements of commercial electrolyzer systems and neither mechanical nor chemical degradation mechanisms are completely understood. The mechanical strength of such membranes can be greatly improved by introducing reinforcements, whereas the chemical stability can be enhanced by adding inorganic or organic fillers like e.g. cross-linking agents or radical scavengers. Raising the operating temperature commonly induces significant thermal degradation to the polymer membranes, including Nafion. The operating temperature can be raised by either introducing additives like e.g. ZrO_2. TiO_2, SiO_2 to common polymer electrolyte matrices (Goñi-Urtiaga et al. 2012) or by alternative polymers or co-polymers such as e.g. poly[2,2'-(m-phenylene)-5,5'-bibenzimidazole] (PBI) (Aili et al. 2011). The additive approaches currently allow operation temperatures in the range of 100–150 °C, whereas PBI can be used at temperatures of up to 200 °C.

Typical catalysts for PEMEC electrolyzers are made of platinum group metals (PGM). On the one hand they show excellent electrochemical activity and stability but on the other hand they are scarce and quite expensive (Carmo et al. 2013). Pt and Pt-Pd mixtures are the most commonly used cathode catalyst materials. Though RuO_2 is known as the most active catalyst for anodic oxygen evolution, it is unstable at higher potentials. Therefore RuO_2 has to be stabilized by Ir or IrO_2, which represents the most widely-used oxygen evolution catalyst system to date (Carmo et al. 2013). Further catalyst stabilization and casually a reduction of the PGM content is realized by doping with non-noble metal-oxides like e.g. SnO_2, Ta_2O_5, Nb_2O_5 (Rasten et al. 2003). Another common method to reduce the PGM-catalyst loading and therefore costs is to disperse catalyst nano-particles onto conducting or non-conducting supporting materials. At the cathode side, carbon based supporting materials like e.g. graphite or carbon black, are widely used. Due to the harsh oxidative conditions at the anode side, those carbon based materials degrade too fast and therefore alternative supporting materials like e.g. TiO_2, TaC or SiC have been developed (Ito et al. 2011; Polonsky et al. 2012). Typical PGM contents are currently in the range of <1 mg/cm^2 at the cathode side and about 2–3 mg/cm^2 at the anode side respectively. Recent developments showed that nano-scale supporting materials such as carbon nanotubes or nano-scale organic whiskers, known as 3 M's nanostructured thin film (NSTF) catalyst technology (Debe 2012), allow a further reduction of the PGM content down to about 0.1 mg/cm^2. Despite reducing PGM metal loadings, alternative catalyst materials based on more abundant elements are gaining growing attention these days. This is currently especially true for hydrogen evolution catalysts, although the oxygen evolution reaction is limiting the electrolyzer performance more significantly. Despite various binary, ternary or quaternary alloys, macrocycles like e.g. Co, Ni-glyoximes or Polyoxometallates like e.g. α-$H_4SiW_{12}O_{40}$ (Millet et al. 2010) have been successfully tested and showed promising initial performance at typical current densities and cell voltages. The implications of alternative catalyst materials on lifetime remain widely unknown. Corresponding evaluations are currently the subject of various research programs.

Bipolar plates (Jung et al. 2009) and current collectors (Grigoriev et al. 2009) are usually made of Ti. The required corrosion resistance is established by a passivating oxide layer. Since that oxide layer negatively influences the contact resistance, the titanium surface has to be protected by surface modifications or protective coatings like Au, Pt, carbides or nitrides. Ta coated metals are seldom used for such cell components as well. Bipolar plates usually contain flow field patterns for optimizing the mass transport properties through the stack. In some cases specially designed current collectors are used instead of such flow fields. Typical current collectors are manufactured in the form of sintered porous media, meshes or felts. Current R&D activities aim at replacing Ti by lower cost materials like e.g. copper, stainless steel or graphite, which necessarily have to be coated with a highly conducting and corrosion resistant protection layer.

3.6.4 Technology Status and Challenges

The PEMEC technology is generally less mature compared to the AEC technology and up to now it has been used exclusively for small scale applications. However, this technology has received great attention in the past decade. This is mainly attributed to its key advantages like high cell efficiencies, high current densities at low corresponding cell voltages hence high power densities and the ability to provide highly compressed hydrogen. Furthermore the PEM technology allows a highly flexible mode of operation enabled by very fast shut-down and start-up times, very fast load followings plus a partial load range of 5–100 %. Those advantages perfectly match many of the basic requirements of Power-to-Gas applications, being directly coupled to fluctuating renewables and being connected to high pressure hydrogen storage units.

The main weak points of the PEM technology are the difficult up-scaling procedures due to the rather high system complexity, the limited global availability of PGMs and the expensive component materials, hence rather high specific system costs. In the past, also low system durabilities have often been noted as disadvantages. Recently significantly improved degradation rates in the range of 10 μV/h or lower have been announced by various manufacturers. This shows that efforts solving stability problems are on the way to catch up with AEC technology. In spite of difficult up-scaling procedures, the system size increased significantly over the past recent years. Major PEM manufactures announced in 2013 that they are working on stacks in the several 100s kW to even MW range, being launched in the upcoming few years.

Considering current R&D trends, it is generally not expected that the cell efficiencies, operating pressures or the current densities will be significantly increased in the near future. It seems that the focus is presently more on further development of upscaling procedures and new stack-component materials, as described in the previous section.

The specific investment costs are currently at least 2 times higher compared to AEC systems. For a serious direct comparison between those technologies one has to take the various differences like e.g. available system sizes, pressure levels and technology status into account. However, in contrast to AEC technology, the cost reduction potential for PEMEC systems is higher. Lowering of costs will be driven by a combination of economy of scale effects and new material and process developments rather than by scale effects on its own. The cost reduction potential, broken down into contributions of each of the stack components, can be summarized as follows. Despite labor costs, bipolar plates and current collectors are the most expensive cell components being responsible for about 50 % of the stack-costs. The MEA accounts for around 25 % of the stack costs (thereof roughly 60 % for the electrolyte membrane and 40 % on PGM catalysts) (Ayers et al. 2011). Because the majority of efficiency losses in electrolyzers are due to slow oxygen evolution reaction kinetics and membrane resistances, those two components are of particular cost-wise and also performance-wise importance.

All this said, it can be concluded that PEM electrolyzers are on the way to becoming a serious competitor to alkaline technology for highly flexible, large scale hydrogen production applications at high power densities. Its modes of operation are highly compatible to Power-to-Gas applications.

3.7 Solid Oxide Electrolyte Electrolysis

3.7.1 Working and Design Principles

SOEC electrolyzers represent the least mature water electrolysis technology of all three discussed here. The general trend of growing interest in water electrolysis technologies and the particular opportunity to significantly reduce the electrical energy demand for electrolysis processes compared to conventional technologies are the main reasons for the recent, significant growing interest in SOEC technology.

A schematic diagram of a SOEC cell is depicted in Fig. 3.4. In SOEC cells a thin, dense solid oxide layer, which becomes conductive for (commonly oxygen-) ions at elevated temperatures, is used as the electrolyte. At both sides of such an electrolyte, porous electrode layers adjoined by current collectors are directly attached. Water (vapor) is usually fed at the cathode side, where the reduction reaction according to Eq. 3.13 takes place. The generated oxygen ions migrate to the anode side, where elemental oxygen evolves according to half-cell reaction 3.14.

$$\text{Cathode}\quad H_2O + 2e^- \rightarrow H_2 + O^{2-} \tag{3.13}$$

$$\text{Anode}\quad O^{2-} \rightarrow 1/2\,O_2 + 2e^- \tag{3.14}$$

Fig. 3.4 Schematic of the operating principle of a SOEC cell

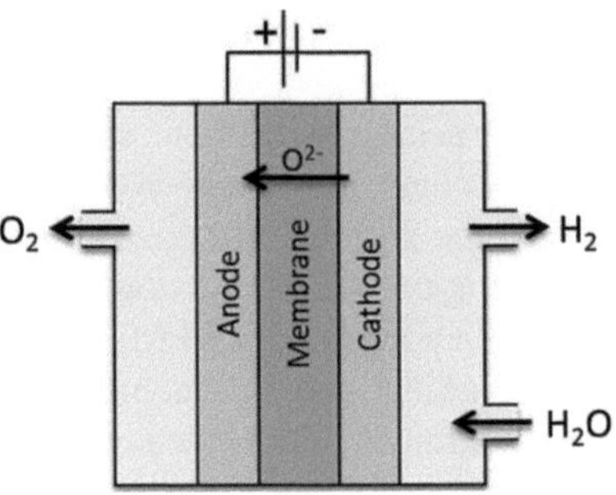

The thickness of each layer mainly depends on its intrinsic conductivity and if it has to provide mechanical support to the cell. Supporting layers are usually the thicker ones used with typical thicknesses in the range of a few 100 μm, where the remaining layers are around 10–30 μm thick.

The single cells can have fairly different geometries and can be carried out either in planar or tubular configuration. Tubular based systems exhibit higher mechanical strength and shorter start-up respectively shut-down times compared to planar ones. Anyhow, planar configurations are currently more widespread due to higher electrochemical performance and better manufacturability.

3.7.2 *Operating Conditions, Performance and Capacities*

SOEC systems are typically operated in a temperature range of 700–1,000 °C (mid to high temperature range). Such high operating temperatures are beneficial for thermodynamic (low electrolysis voltages) as well as for kinetic (non PGM catalysts, low overvoltages) reasons. The downside is that such high temperatures cause severe problems with fast degradation of the cell components. Therefore, a significant part of current research efforts aims at systems being operated at lower temperatures like 500–700 °C (low to mid temperature range). The current densities could be principally as high as those known from PEMEC systems but due to strong degradation the practical current densities are kept more in the AEC-range, namely 300–600 mA/cm². The typical corresponding cell voltages are around 1.2–1.3 V, which result in remarkably low electricity consumptions of less than 3.2 kWh$_{el}$/scm H₂. The particular electricity consumption depends on the available heat sources. The energy for water splitting can be either fully supplied by electrical power (autothermal) or partly supplied by an external high temperature heat source (allothermal), which causes variations of around 0.6 kWh/scm. If steam is not available as a feedstock, an additional low temperature heat source is required, what further increases the energy consumption of the electrolyzer. The system efficiencies are typically in the range of 90+ %, taking the energy demands for heat and electricity into account.

SOEC systems are presently mainly operated at atmospheric pressure. According to the general trend in water electrolysis technologies towards higher system pressures, some SOEC systems pressured up to 25 bar have been developed (Jensen et al. 2010). With a proper temperature control, a highly dynamic operation is possible.

Degradation rates reported in the literature vary greatly from a few percent up to 20 % per 1,000 h of operation. These huge variations can be explained by completely different set of materials used and quite differing operating conditions.

3.7.3 Cell Components

The core SOEC components are usually made of ceramic materials. Due to the high operating temperatures, the stability of their phases respectively their morphologies are of particular importance. Additionally the individual thermal expansion coefficients of each of the layers should match as close as possible, in order to prevent cracking of the thin ceramic layers upon temperature variations. The tuning of particular material parameters in order to satisfy the needs for certain levels of stability, conductivity, expansion coefficients, and so on... is usually done by doping.

Yttria-stabilized zirconia (YSZ) is currently the most widely used electrolyte for SOEC systems working at high operating temperatures (Ni et al. 2008). For mid temperature levels ceria based ceramics like Sm-doped ceria (SDC) or Gd-doped ceria GDC are promising candidates, whereby Sr and Mg doped lanthanum gallates (LSGM) are favored at lower operating temperature ranges due to their still reasonably high conductivities (Laguna-Bercero 2012).

The typical operating temperatures of SOEC electrolyzers are so high, that state-of-the art SOEC electrolyzers do not necessarily need PGM catalysts for reasonably active catalytic zones. However, precious metals are often used for thin, electrical contact layers.

Cathode materials commonly contain Ni. Although Ni actively induces hydrogen evolution, it only conducts electrons limiting the reaction zone to the cathode electrolyte interface. To extend that zone, Ni is usually mixed with ion conducting particles similar to or the same as the electrolyte material. This mixture, called cermet, is currently the standard type of cathodes consisting mainly of Ni/YSZ or Ni/SDC (Ni et al. 2008).

Most common anode materials are composite electrodes of YSZ with perowskite type mixed oxides like Sr-doped strontium-doped lanthanum manganite (LSM), Sr-doped lanthanum cobalt oxide (LSC) or Sr-doped lanthanum ferrite (LSF) Lanthanum strontium cobalt ferrite LSCF (Ni et al. 2008).

Depending on the particular set of materials, it is possible that the electrolyte material reacts with components of the attached electrodes. Therefore protecting oxide interlayers are often additionally applied between those layers.

The interconnects between the adjoined single cells are typically made of ceramic materials and for low and mid temperature applications also metallic. Sealing materials can be either glasses, glass ceramics or glass composite seals (Menzler et al. 2010).

In conclusion, there is a lot of fundamental research currently in the area of SOEC systems. The research has to be further conducted intensively in order to obtain sufficiently improved SOEC component materials, which are able to serve commercial electrolyzer needs.

3.7.4 Technology Status and Challenges

To date, the SOEC technology is the least developed among the main water electrolysis technologies. Currently, most operated SOEC units are on a laboratory scale with maximum power levels in the lower kW range. Significant improvements of the state-of-the-art system components are required prior to commercialization of SOEC electrolyzers. The most critical issues are the high degradation rates, which can be resolved by stabilizing the particular existing component materials, by developing new materials or by lowering the operation temperature to 500–700 °C. All three strategies are currently more or less equally researched.

However, the SOEC technology offers a couple of unique properties as discussed in the following. The efficiency potential is generally significantly higher compared to low temperature electrolysis technologies. External high temperature heat sources allow a further reduction of the electrical energy demand for proper SOEC systems below 3 kWh_{el}/scm H_2. Therefore the overall energy costs for hydrogen production are typically lower compared to low temperature electrolysis technologies, since a kWh of heat is usually significantly cheaper compared to a kWh of electricity.

Because of their operation temperature, SOEC systems are highly reversible devices, which can also be operated in reverse mode acting as a fuel cell in a single device (Ruiz-morales and Marrero-lo 2011). So called unitized reversible fuel cells (URFCs) are a kind of light-weight rechargeable batteries offering many commercially interesting opportunities for weight-critical applications. In fact, SOEC cells are often derived from solid oxide fuel cells (SOFC) with no or minor modifications. Although the electrolysis mode adds some additional requirements, especially at the oxygen evolution electrode, the SOEC technology widely benefit from the actively driven SOFC developments.

Another interesting feature is that SOEC devices can be used to electrochemically reduce CO_2 to CO instead of producing hydrogen. Commonly this feature is utilized in a co-electrolysis process (Ebbesen et al. 2012), where H_2O/CO_2 are jointly reduced to a mixture of H_2 and CO, called syngas. This offers the basis for a wide range of synthetic products like various fuels, fertilizers, solvents and synthetic materials.

In summary, the commercialization of SOEC systems will still take a while but they are highly efficient and interesting electrolysis systems for various kinds of applications, which are not only limited to hydrogen production.

3.8 Conclusion

Currently there are three main water electrolysis technologies available, namely alkaline electrolysis (AEC), polymer electrolyte membrane electrolysis (PEMEC), and solid oxide electrolyte electrolysis (SOEC). Each of them remains at a different level of development and to date only AEC and PEMEC systems are commercially available.

The main technical differences between these three technologies are the operating temperature, the operating current density respectively voltage, the class of materials used for catalysis, the pH value and the type of the electrolyte used and thus the configuration of the particular electrolyzer systems. An overview of the important parameters of the three main water electrolysis technologies is given in Table 3.1. For each parameter typical values are presented.

The high temperature electrolysis technology (SOEC), operating at 700–1,000 °C, has the highest efficiency potential. It is currently the least developed technology and suffers from severe material degradation issues. A lot of fundamental research has to be carried out to overcome those limitations. The alkaline, low temperature electrolysis technology is the oldest, currently most mature and cheapest technology available. In large-scale electrolytic hydrogen production plants, alkaline electrolyzers are used exclusively so far. However, low current densities and rather limited modes of dynamic operation are currently major limitations of that technology. To make the AEC technology more compatible to Power-to-Gas applications, further developments are essential. The acidic solid polymer electrolyte (PEMEC)

Table 3.1 Important parameters of the main water electrolysis technologies

	AEC	PEMEC	SOEC
Ions electrolyte	OH^-	H^+	O^{2-}
Current density (A/cm^2)	<0.5	>1	<0.3
Cell voltage (V)	>1.9	>1.8	>1
Temperature (°C)	60–80	60–80	700–1,000
Operating pressure (bar)	<30	<200	<25
(Voltage) Efficiency (%)	60–80	65–80	–
Spec. el. energy consumption (kWh/scbm)	>4.6	>4.8	<3.2
Lower partial load range [% of nominal load (NL)]	30–40	0–10	–
Overload (% of NL)	<150	<200	–
Capacity (scbm H$_2$)	<760	<40	<5
Cell area (m^2)	<4	<0.3	<0.01

scbm standard cubicmeter

technology has made significant progress over the past century and is on its way to leave niche applications. Due to various unique advantages over alkaline systems like the compact system design, high current densities, high operating pressures, high flexibility with respect to modes of operation and wide partial load ranges, the PEMEC technology offers a great potential to become a serious competitor to alkaline electrolysis systems for many types of applications. Due to these advantages the PEMEC technology is probably the most compatible technology for Power-to-Gas applications at present. The most limiting disadvantages of that technology are its high costs, the limited resources and missing adequate scale up procedures.

Independent from the particular water electrolysis technology, the major drawbacks are the limited capacities of currently available electrolyzers, suboptimal degradation behaviors and high investments respectively operating costs of electrolyzer systems. Substantial R&D efforts are still necessary for each of the water electrolysis technologies to overcome those problems and to pave the way for a broader implementation of electrolytic hydrogen production into the market.

References

Aili D, Hansen MK, Pan C et al (2011) Phosphoric acid doped membranes based on Nafion®, PBI and their blends—membrane preparation, characterization and steam electrolysis testing. Int J Hydrogen Energy 36:6985–6993. doi:10.1016/j.ijhydene.2011.03.058

Ayers KE, Capuano C, Anderson EB (2011) Recent advances in cell costs and efficiency for PEM based water electrolysis. Electrochem Soc 2013

Carmo M, Fritz DL, Mergel J, Stolten D (2013) A comprehensive review on PEM water electrolysis. Int J Hydrogen Energy. doi: 10.1016/j.ijhydene.2013.01.151

De Levie R (1999) The electrolysis of water. J Electroanal Chem 476:92–93

Debe MK (2012) Electrocatalyst approaches and challenges for automotive fuel cells. Nature 486:43–51. doi:10.1038/nature11115

Dorf RC (ed) (2004) CRC—handbook of engineering tables, 2nd edn. CRC Press LLC, Boca Raton

Ebbesen SD, Knibbe R, Mogensen M (2012) Co-electrolysis of steam and carbon dioxide in solid oxide cells. J Electrochem Soc 159:F482–F489. doi:10.1149/2.076208jes

Goñi-Urtiaga A, Presvytes D, Scott K (2012) Solid acids as electrolyte materials for proton exchange membrane (PEM) electrolysis: review. Int J Hydrogen Energy 37:3358–3372. doi:10.1016/j.ijhydene.2011.09.152

Grigoriev SA, Millet P, Volobuev SA, Fateev VN (2009) Optimization of porous current collectors for PEM water electrolysers. Int J Hydrogen Energy 34:4968–4973. doi:10.1016/j.ijhydene.2008.11.056

Holladay JD, Hu J, King DL, Wang Y (2009) An overview of hydrogen production technologies. Catal Today 139:244–260. doi:10.1016/j.cattod.2008.08.039

Ito H, Maeda T, Nakano A, Takenaka H (2011) Properties of Nafion membranes under PEM water electrolysis conditions. Int J Hydrogen Energy 36:10527–10540. doi:10.1016/j.ijhydene.2011.05.127

Jensen SH, Sun X, Ebbesen SD et al (2010) Hydrogen and synthetic fuel production using pressurized solid oxide electrolysis cells. Int J Hydrogen Energy 35:9544–9549. doi:10.1016/j.ijhydene.2010.06.065

Jung H-Y, Huang S-Y, Ganesan P, Popov BN (2009) Performance of gold-coated titanium bipolar plates in unitized regenerative fuel cell operation. J Power Sources 194:972–975. doi:10.1016/j.jpowsour.2009.06.030

Kaninski MPM, Nikolic VM, Tasic GS, Rakocevic ZL (2009) Electrocatalytic activation of Ni electrode for hydrogen production by electrodeposition of Co and V species. Int J Hydrogen Energy 34:703–709. doi:10.1016/j.ijhydene.2008.09.024

Kreuter W, Hofmann H (1998) Electrolysis: the important energy transformer in a world of sustainable energy. Int J Hydrogen Energy 23:661–666

Laguna-Bercero MA (2012) Recent advances in high temperature electrolysis using solid oxide fuel cells: a review. J Power Sources 203:4–16. doi:10.1016/j.jpowsour.2011.12.019

Lal B, Raghunandan M, Gupta M, Singh R (2005) Electrocatalytic properties of perovskite-type obtained by a novel stearic acid sol–gel method for electrocatalysis of evolution in KOH solutions. Int J Hydrogen Energy 30:723–729. doi:10.1016/j.ijhydene.2004.07.002

Marini S, Salvi P, Nelli P et al (2012) Advanced alkaline water electrolysis. Electrochim Acta 82:384–391. doi:10.1016/j.electacta.2012.05.011

Menzler NH, Tietz F, Uhlenbruck S et al (2010) Materials and manufacturing technologies for solid oxide fuel cells. J Mater Sci 45:3109–3135. doi:10.1007/s10853-010-4279-9

Millet P, Ngameni R, Grigoriev SA et al (2010) PEM water electrolyzers: from electrocatalysis to stack development. Int J Hydrogen Energy 35:5043–5052. doi:10.1016/j.ijhydene.2009.09.015

Mittelsteadt CK, Staser JA, Systems GE (2012) Electrolyzer membranes. Polym Sci Compr Ref 10:849–871. doi:10.1016/B978-0-444-53349-4.00296-X

Modica G, Maffi S, Montoneri E et al (1986) Aromatic polymers for advanced alkaline water electrolysis—III. Polysulphone-TiO$_2$ films. Int J Hydrogen Energy 11:307–308

Ni M, Leung M, Leung D (2008) Technological development of hydrogen production by solid oxide electrolyzer cell (SOEC). Int J Hydrogen Energy 33:2337–2354. doi:10.1016/j.ijhydene.2008.02.048

Otero J, Sese J, Michaus I et al (2014) Sulphonated polyether ether ketone diaphragms used in commercial scale alkaline water electrolysis. J Power Sources 247:967–974. doi:10.1016/j.jpowsour.2013.09.062

Pletcher D, Li X (2011) Prospects for alkaline zero gap water electrolysers for hydrogen production. Int J Hydrogen Energy 36:15089–15104. doi:10.1016/j.ijhydene.2011.08.080

Pletcher D, Li X, Wang S (2012) A comparison of cathodes for zero gap alkaline water electrolysers for hydrogen production. Int J Hydrogen Energy 37:7429–7435. doi:10.1016/j.ijhydene.2012.02.013

Polonsky J, Paidar M, Bouzek K, Mazu P (2012) Non-conductive TiO$_2$ as the anode catalyst support for PEM water electrolysis. doi:10.1016/j.ijhydene.2012.05.129

Rasten E, Hagen G, Tunold R (2003) Electrocatalysis in water electrolysis with solid polymer electrolyte. Electrochim Acta 48:3945–3952. doi:10.1016/j.electacta.2003.04.001

Ruiz-morales JC, Marrero-lo D (2011) RSC advances symmetric and reversible solid oxide fuel cells. R Soc Chem 1:1403–1414. doi:10.1039/c1ra00284h

Singh RN, Mishra D, Sinha ASK, Singh A (2007) Novel electrocatalysts for generating oxygen from alkaline water electrolysis. Electrochem commun 9:1369–1373. doi: 10.1016/j.elecom.2007.01.044

Subbaraman R, Tripkovic D, Chang K-C et al (2012) Trends in activity for the water electrolyser reactions on 3d M(Ni Co, Fe, Mn) hydr(oxy)oxide catalysts. Nat Mater 11:550–557. doi:10.1038/nmat3313

Suntivich J (2011) A perovskite oxide optimized for oxygen evolution catalysis from molecular orbital principles. Science 334:1383–1385. doi: 10.1126/science.1212858

Trasatti S (1999) Water electrolysis: who first? J Electroanal Chem 476:90–91

Vermeiren P (1998) Evaluation of the Zirfon separator for use in alkaline water electrolysis, 3199

Zeng K, Zhang D (2010) Recent progress in alkaline water electrolysis for hydrogen production and applications. Prog Energy Combust Sci 36:307–326. doi:10.1016/j.pecs.2009.11.002

Chapter 4
Methanation

Methanation describes the heterogeneous, gas-catalytic or biological synthesis of CH_4 from H_2 and CO/CO_2 or in case of the biological path, alternatively from other carbon sources. It is the second substantial, but optional process step beside the electrolysis within the Power-to-Gas concept. The chemism of the methanation reaction is known for more than one century and chemical methanation processes have been state of the art for several decades. They have been and still are applied to produce substitute natural gas (SNG) from synthesis gas derived from coal or biomass. Gas purification in chemical or petrochemical industries is another widely used application of the methanation process. Although methanation is technological mature in these fields of application, specific differences and challenges arise when used as process step within the Power-to-Gas concept. This chapter gives an overview of the state-of-the art of methanation processes and the specifics for the application of this technology as part of Power-to-Gas.

4.1 State of the Art of Methanation Processes

Starting with the chemical process routes followed by a view on the biological route, this chapter gives a general overview of the state-of-the art of methanation processes currently used in industry, but mainly utilized for applications other than Power-to-Gas.

4.1.1 Chemical Fundamentals

The Sabatier reaction was discovered in 1902, and is described by

$$CO_{(g)} + 3H_{2(g)} \leftrightarrow CH_{4(g)} + H_2O_{(g)} \quad \Delta H_R^0 = -206.2\,kJ/mol \quad (4.1)$$

© The Author(s) 2014

M. Lehner et al., *Power-to-Gas: Technology and Business Models*,
SpringerBriefs in Energy, DOI 10.1007/978-3-319-03995-4_4

In combination with the shift conversion

$$CO_{2(g)} + H_{2(g)} \leftrightarrow CO_{(g)} + H_2O_{(g)} \quad \Delta H_R^0 = +41.2\,kJ/mol \tag{4.2}$$

a formulation for the reaction of CO_2 with hydrogen can be given

$$CO_{2(g)} + 4H_{2(g)} \leftrightarrow CH_{4(g)} + 2H_2O_{(g)} \quad \Delta H_R^0 = -165.0\,kJ/mol \tag{4.3}$$

Reaction enthalpies in Eqs. (4.1)–(4.3) are given for a temperature of 25 °C. Equation (4.3) is often interpreted as the sum of Eqs. (4.1) and (4.2), i.e. CO_2 methanation is achieved by the intermediate conversion to CO. The reaction mechanism is still under investigation and not finally clarified. Reactions (4.1) and (4.3) are strongly exothermic, and all are equilibrium reactions. Equilibrium constants and reaction heats for different temperatures are given in Elvers et al. (1989). Lower temperatures result in significantly higher equilibrium constants, and therefore in better conversion rates. But lower temperatures cause unfavorable reaction kinetics, hence appropriate catalysts are used. Due to the volume reduction of the chemical reaction [Eq. (4.3)], higher pressures basically support better conversion rates. Since CO and CO_2 participate in the reaction scheme, depending on the operating conditions, also the Boudouard equilibrium may have to be considered as an undesirable side reaction forming coke in the reactor:

$$CO_{2(g)} + C_{(s)} \leftrightarrow 2CO_{(g)} \quad \Delta H_R^0 = +172,45\,kJ/mol \tag{4.4}$$

The product gas leaving the reactor contains steam, CO and unconverted educts beside the product CH_4. The product composition can be influenced by the methanation process concept, the reaction parameters and also by the reactor types used. Additionally, the applied catalyst influences kinetics, conversion rate and selectivity of the process.

Catalytic active substances for the hydrogenation of CO_2 or CO are group VIII metals, i.e. the Fe-group, the Co-group and the Ni-group (Mills and Steffgen 1974). Mainly due to the reasonable cost and satisfactory performance in terms of conversion rates and selectivity, Ni based catalysts are widely utilized for methanation processes today. Usually, silica based carriers are used, but also zeolites or metal carriers are known (Kaltenmaier 1988; Wang et al. 2011; Weatherbee and Bartholomew 1982). Basically, catalysts are sensitive to poisons, which may result in catalyst deactivation. Typical catalyst poisons are heavy metals, but also sulfur compounds or oxygen (Bartholomew 2001). This is of special significance for methanation processes as part of Power-to-Gas, as described later. Generally valid statements on both the kinetics and the mechanism of the hydrogenation of CO_2 or CO are still not available.

4.1.2 Process Concepts and Stage of Development

In a historic view, mainly two phases of process development can be identified (Kopyscinski et al. 2010): forced by the first oil crisis and by strategic considerations, coal-to-gas (CtG) processes have been developed in the 1970s based on fossil coal as feedstock. The typical process path is gasification, gas cleaning and conditioning followed by the methanation and a necessary gas upgrading to meet the requirements for the injection of the produced substitute natural gas (SNG) into the gas grid. Industrial scale plants based on this technology have been erected and operated in USA (US Department of Energy 2014), and as coal-to-liquid (CtL) plant also by Sasol in South Africa.

The second phase, starting around the year 2000, focuses on the conversion of biomass as feedstock (biomass-to-gas/BtG or biomass-to-liquid/BtL). Both smaller plant scales and differing feed gas compositions of synthesis gas derived from biomass, make the direct utilization of the previously developed plant concepts for coal-to-gas plants difficult or impossible to apply. Hence, new process developments have been initialized. The renaissance of methanation is mainly driven by the intended transition of the energy system towards renewable sources, and also by rising prices for natural gas. Within the process chains of CtG and BtG, methanation is one process step. The developed chemical methanation processes of the past decades can be classified as follows (Bajohr et al. 2011):

- 2-phase systems (gaseous educts, solid catalyst):
  – Fixed bed
  – Fluidized bed
  – Coated honeycombs
- 3-phase systems (gaseous educts, liquid heat carrier, solid catalyst)
  – Bubble column (slurry)

A main focus is the heat management of the reactors. As described earlier, all chemical reactions involved are strongly exothermic. Therefore, the temperature regulation of the processes is challenging, and is solved in various ways depending on the reactor type. In Table 4.1 an overview is given of methanation processes and reactor types developed from the 1950s to date. Further information can be found in Elvers et al. (1989, Kopyscinski et al. (2010), Bajohr et al. (2011).

Fixed Bed Methanation utilizes catalysts, in pellet form, some millimeters in size which are dumped randomly into the reactor forming a preferably homogeneous, static catalyst bed. Due to the strongly exothermic reactions, the temperatures of the 250–300 °C preheated gases rise significantly. Depending on the operation pressure, conversion rates and selectivity is decreased with temperatures higher than 400–500 °C. Therefore, fixed bed methanation processes always split the reaction in a cascade of reactors with gas cooling, gas recycling and reaction heat recovery between each reactor step. The temperature control is important for all fixed bed types in order to avoid local temperature peaks in the bed (hot spots) which may result in catalyst destruction. Mass transfer limitations between the gases and the solid catalyst are another disadvantage of the fixed bed types, whilst the mechanical stress to the catalyst is comparably low, and hence one of the main benefits. According to Table 4.1, the Lurgi, TREMP™, Linde, HICOM and RMP processes belong to this reactor type, amongst others. In Fig. 4.1 the basic flow chart of the TREMP™ process (Topsøe's Recycle Energy Efficient Methanation Process; trademark of Haldor Topsøe A/S, Denmark) is depicted (Kopyscinski

Table 4.1 Selection of methanation process developments (1955–2013) (Elvers et al. 1989; Kopyscinski et al. 2010; Bajohr et al. 2011)

Process	Year	Stage of development	Reactor type	No. of stages	Temperature range [°C]	Pressure [bar]	Operating hours	Educts
Lurgi	1974	Commercial	FB	2	~450	>18	Several 1,000	Coal
Comflux	1980 (2008)	Pilot	FL	1	400–500	20–60	Several 1,000	Coal (later biomass)
TREMP	1980	Semi-commercial	FB	3	300–700 (250)	30	Several 1,000	Coal, petrol coke, biomass
SuperMeth/ Conoco-Meth	1979/1974	Pilot/demo	FB	4/4	n.s.	~80	n.s.	Coal
HYGAS	~1955	Pilot	FB	2	280–480	70	n.s.	Coal
HICOM	1981	Pilot	FB	4	230–640	25–70	>15.000 as pilot	Coal
Linde	1979	Semi-commercial (Methanol-synthesis)	FB	2–3	300–750	20	n.s.	n.s.
RMP	1974	Pilot	FB	4–6	315–780	1–70 (4,5–77)	n.s.	Coal, fuel oil
Bi-Gas	1965	Pilot	FL	1	40–530	86 (69–87)	n.s.	Coal
Synthane Project	1970 (until 1980)	Lab	Pipe with Raney Nickel	2	300 (390)	40–50 (20) (70)	<1,000	n.s.
CCG (catalytic coal gasification)	early 1980s	Lab/demo	FL	1	700	30	>2,000 as Demo	n.s.
LPM	1976 (1981)	Pilot	BC	1	~340 (315–360)	~70 (34–53)	n.s.	n.s.
Hydro-methanation (bluegas)	Ongoing	Pilot	FL	1	600–700	n.s.	~1,000	n.s.

(continued)

Table 4.1 (continued)

Process	Year	Stage of development	Reactor type	No. of stages	Temperature range [°C]	Pressure [bar]	Operating hours	Educts
Hydrogasification process	2009	Lab	Direct Gasification C/H_2	1	870	70	n.s..	Coal
AER (ZSW)	Ongoing	Lab	n.s.	n.s.	250–500	6,5	<1,000	Biomass
PSI	Ongoing	Pilot/demo	FL (Comflux)	1	400–500	20–60	<1,000	Biomass
Bio-SNG (Güssing, Austria)	2006	Pilot/demo	FL (Comflux)	1	350	2–5	>1,000	Biomass
GoBiGas	2013	Demo	FB (TREMP)	2	300–700	25	Start-up in 2014	Biomass

FB: Fixed bed; *FL*: Fluidized bed; *PFR*: Plug flow reactor; *BC*: Bubble column; *Pilot*: Pilotplant; *Lab*: Laboratory scale; *Demo*: Demonstration plant; *n.s.*: not specified

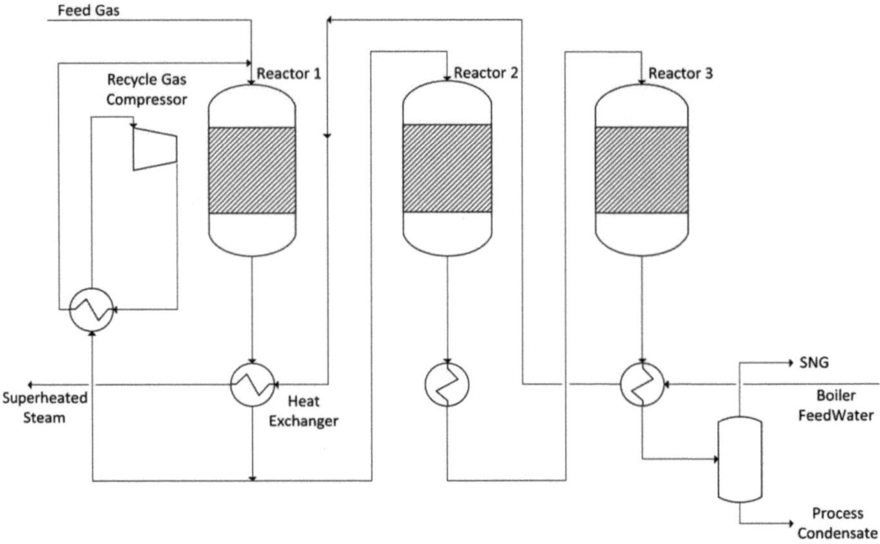

Fig. 4.1 Example for a fixed bed methanation: Haldor Topsøe TREMP™ process (Kopyscinski et al. 2010; Haldor Topsøe 2009)

et al. 2010; Haldor Topsøe 2009), which has been developed in the 1970s and 1980s as a cycle process to store and to distribute process heat from nuclear reactors. Recently, this process concept has been adopted for a commercial biomass gasification project at Sweden [GoBiGas, Table 4.1 and (GoBiGas 2014)].

Lurgi developed a methanation process with two adiabatic fixed bed reactors and internal gas recycling. Two pilot plants have been erected: one at Sasolburg, South Africa, and a second at Schwechat, Austria. At the Sasolburg plant, a side stream of the Fischer-Tropsch-synthesis has been utilized as feed gas for the methanation. In the second pilot plant, naphtha has been converted to methane. In 1984, the Lurgi process concept was realized in an industrial scale at the Great Plains Synfuels Plant, North Dakota, USA (US Department of Energy 2014). As shown in Fig. 4.2, lignite coal (18.000 t/d) is used as feedstock for the gasifier. The gas conditioning is complex, and consists of a gas cooling, a shift conversion reactor and a Rectisol unit (gas scrubbing by cryogenic methanol). The actual Lurgi fixed bed methanation is at the end of the process chain. The side product stream of CO_2 is used for enhanced oil recovery (EOR). The average availability of the plant is 98.7 % with a production rate of 4.81×10^6 m^3 SNG/day. Average catalyst life time is 4 years (US Department of Energy 2014).

Fluidized Bed Methanation is characterized by an approximately isothermal temperature profile in the reactor which is achieved by strong turbulence as a result of the fluidization of the solid catalyst particles. The necessary force for fluidization is applied by the gas. Hence, the operating range of a fluidized bed is limited to a certain gas flow range which results in limitations of unsteady operation. Furthermore, the movement of the catalyst particles in the fluidized

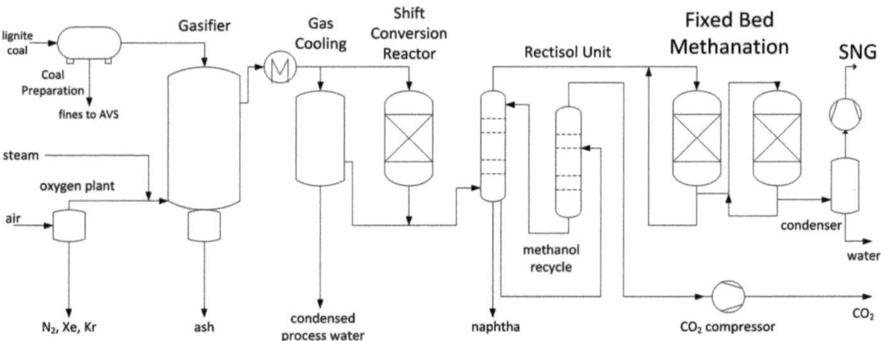

Fig. 4.2 Simplified process flow chart of great plains synfuels plant, modified from Kopyscinski et al. (2010)

bed generates abrasion, both at the catalyst and the reactor internals. The main advantages of this reactor concept are a good heat release and a high specific surface area of the catalyst combined with reduced mass transfer limitations. Consequently, reactor cascades are avoided, and so a simplified set-up is realized compared with fixed bed systems (Fig. 4.3).

Examples of fluidized bed methanation are the Comflux and Bi-Gas concept, see Table 4.1. The Comflux process has been developed by Thyssengas and the University of Karlsruhe for the production of SNG from synthesis gas derived from coal. A pilot plant has been operated at Ruhrchemie Oberhausen with a capacity of 2000 m^3 SNG/h (20 MW$_{SNG}$). The project was stopped in the mid 1980s due to declining oil prices. The Comflux process concept was revitalized by the Paul Scherrer-Institut (PSI, Switzerland) which performed experiments with a 10 kW$_{SNG}$ unit end of 2004 (Kopyscinski et al. 2010). The focus was to hydrogenate a synthesis gas derived from biomass. Rapid catalyst deactivation has been observed caused by organic sulfur species. With improved desulfurization of the feed gas to the methanation, catalyst lifetime has been prolonged significantly (Seemann et al. 2004). The fluidized bed process was successfully up-scaled to a 1 MW$_{SNG}$ plant at Güssing, Austria, and is in full operation since end of 2009 (Biollaz et al. 2009).

Bubble Columns operate the methanation process in a 3-phase system: gaseous educts, solid catalyst and, additionally, a liquid heat carrier medium. Originally, the catalytic liquid phase methanation was developed by Chem System Inc. (USA) in the 1970s (see Table 4.1, Fig. 4.4 and Kopyscinski et al. 2010; Bajohr et al. 2011; Chem Systems Inc. 1979). By introducing a liquid phase, the heat release of the exothermic reactions is promoted and thus an isothermal temperature profile in the reactor is achieved. Furthermore, catalyst abrasion is reduced compared to the fluidized bed. The hydraulic operation of a three-phase bubble column is quite sophisticated. Due to the introduction of the liquid phase, an additional mass transfer resistance between the gaseous educts and the solid catalyst incurs which may influence negatively the kinetics of the total process.

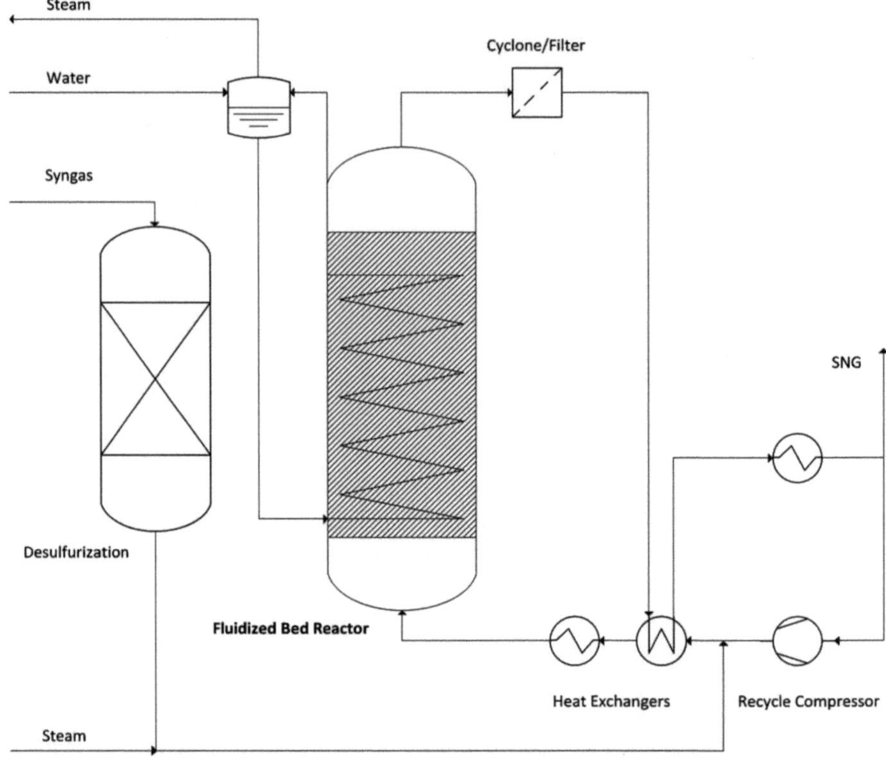

Fig. 4.3 Comflux (Thyssengas) process, modified from Kopyscinski et al. (2010)

Chem Systems utilized a mineral oil as liquid heat carrier medium, and observed degradation of the mineral oil due to reduced temperature stability. The project was stopped in 1981 (Bajohr et al. 2011).

The process concept of liquid phase methanation has been revitalized by the Forschungszentrum Karlsruhe and DVGW, Germany (Bajohr et al. 2011). This new development utilizes ionic liquids instead of mineral oils in order to overcome the observed problems at the Chem Systems process. It aims to solve problems of part loads to the methanation reactor, as well as of modularization of the reactor design. Both requirements arise from the specifics of methanation's application to Power-to-Gas systems, which will be described later.

4.1.3 Biological Process Routes

The above described chemical catalysts and process routes can be substituted by bio-catalysts (enzymes) where the methanation of hydrogen and carbon dioxide is carried out in a biological system. Methanogenic bacteria, which belong to the

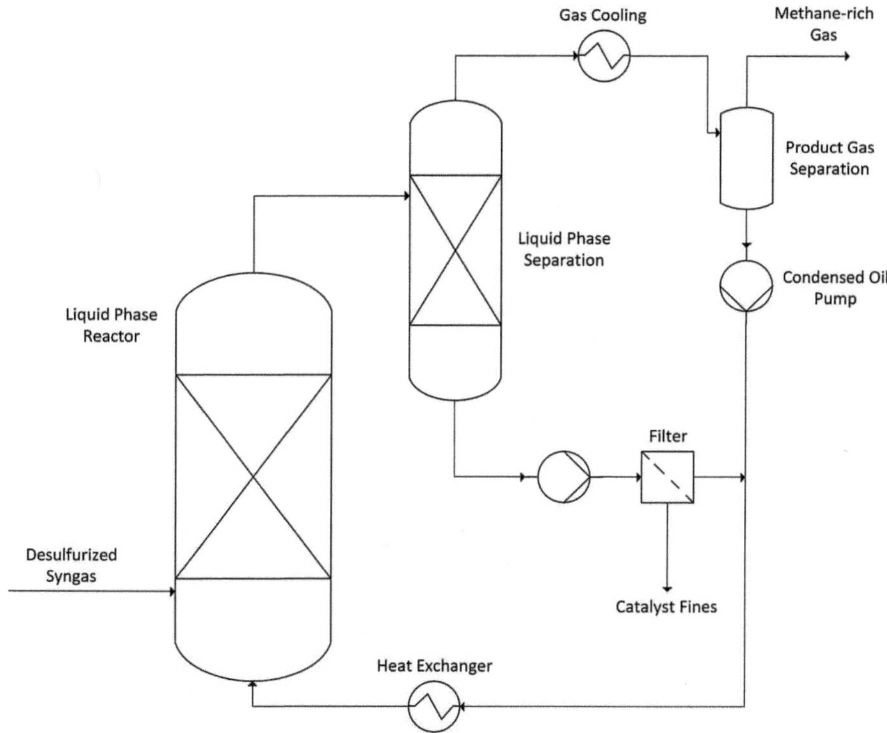

Fig. 4.4 Liquid phase methanation concept Chem Systems Inc. (1979), modified from Kopyscinski et al. (2010)

domain of Archaea, produce the necessary enzymes. Biological methanation is particularly known in biogas processes in which two main reaction paths can be distinguished:

Acetoclastic methanogenesis

$$CH_3COOH_{(g)} \leftrightarrow CH_{4(g)} + CO_{2(g)} \quad \Delta G_R^0 = -33.0 \, \frac{kJ}{mol} \tag{4.5}$$

and the hydrogenotrophic methanogenesis

$$CO_{2(g)} + 4H_{2(g)} \leftrightarrow CH_{4(g)} + 2H_2O_{(g)} \quad \Delta G_R^0 = -135.0 \, kJ/mol \tag{4.6}$$

Equation (4.6) is equivalent to Eq. (4.3). Both metabolic pathways are catalyzed by different microbes which all belong to the domain of Archaea. The methane production based on acids (Eq. (4.5)) is the dominant process route for the decomposition of biomass. But also the second biological pathway (Eq. 4.6) is utilized in a biogas plant settled with a mixed microbe population (Karakashev et al. 2005).

Different process concepts are available for the biological catalysis of hydrogen to methane. Either an optimized biogas plant is utilized (integrative methanation), in which both described pathways occur simultaneously, or in reactors for the selective hydrogen utilization (selective methanation). The integrative methanation is described in literature both in laboratory and pilot scale (Luo et al. 2012).[1] Hydrogen is used as co-substrate in addition to manure or sewage sludge. Hydrogen conversion rates of 80 % are reported which depend on the hydrogen partial pressure and the mixing intensity. The control of the pH value in the system and an instantaneous conversion of the hydrogen to methane seem to be crucial for a stable operation of the process. The selective methanation utilizes adapted microbes under optimized process conditions in a bioreactor. It can be linked to a biogas process, but a self-sustaining operation is also possible which then needs beside the hydrogen an own carbon source. Laboratory tests indicate a hydrogen conversion rate greater than 90 % with operating temperatures of 55 °C (thermophilic). The gas-liquid mass transfer seems to be the limitative factor (Luo and Angelidaki 2012). Among others, Krajete GmbH, Austria, commercialize such systems.[2]

Biological methanation is an upcoming technology gaining more and more importance. The advantages compared with the conventional, chemical methanation are operation at moderate temperatures (30–60 °C) and atmospheric pressure, as well as a higher tolerance against pollutant substances in the feed gases. One of the disadvantages is the operation in a three-phase system resulting in a mass transfer limitation between the gas and the liquid phase. Microbes are creatures which need, beside the described feed gases, also other nutrients like salts which have to be provided in the bioreactors. The long-term stability of such biological systems and the microbes itself, the selectivity of the biological reactions as well as the performance under intermittent operating conditions are still subject of research.

Table 4.2 summarizes the properties of the introduced methanation concepts. It is obvious that all of the process concepts show specific advantages and

Table 4.2 Comparison of different methanation concepts, modified and extended from Bajohr et al. (2011)

Concept	Chemical methanation			Biological methanation
	Fixed bed	Fluidized bed	Bubble column	
Heat release	− −	+	++	++ (no issue)
Heat control	− −	o	++	++ (no issue)
Mass transfer	o	++	− −	− −
Kinetics	+	+	+	o
Load flexibility	o	− −	o	− −
Stress on catalyst	+	− −	+	++ (no issue)

++ *very good,* + *good, o average,* − *poor,* − − *very poor*

---

[1] http://www.viessmann.de/content/dam/internet-global/pdf_documents/koeb_mawera/MicrobEnergy_power-to-gas.pdf. Accessed 6 April 2014.

[2] http://www.krajete.com/en/. Accessed 6 April 2014.

disadvantages. Experience on an industrial scale is only available in case of the fixed bed chemical methanation. In any case, the specific requirements for utilizing methanation as Part of Gower-to-Gas demand further research and development, as described in the following chapter.

4.2 Methanation as Part of Power-to-Gas

Although the primary objective of the methanation process, the conversion of hydrogen and carbon (di)oxide to methane, remains unchanged, a number of specific differences arise when methanation is adopted to the Power-to-Gas concept. A schematic overview of a chemical methanation unit of a Power-to-Gas plant is given in Fig. 4.5.

The electrolysis unit supplies the necessary hydrogen for the methanation process. The operation of the electrolysis is unsteady following the fluctuating input of the renewable power to the system. Particularly chemical methanation has to be steadily operated with elevated temperatures and pressures (see Table 4.1). Neither frequent start-up and shut-down cycles can be realized, nor significant load changes. The sensitivity to load changes depends on the reactor concept, but basically the load flexibility is limited (Table 4.2). Therefore, an intermittent storage of hydrogen is necessary where the size of the storage tank both depends on the fluctuations in electrolysis and the load flexibility of methanation (Schaub et al. 2014). The same considerations are basically also valid for the second educt gas, carbon dioxide. The intermittent storage of carbon dioxide is simpler compared

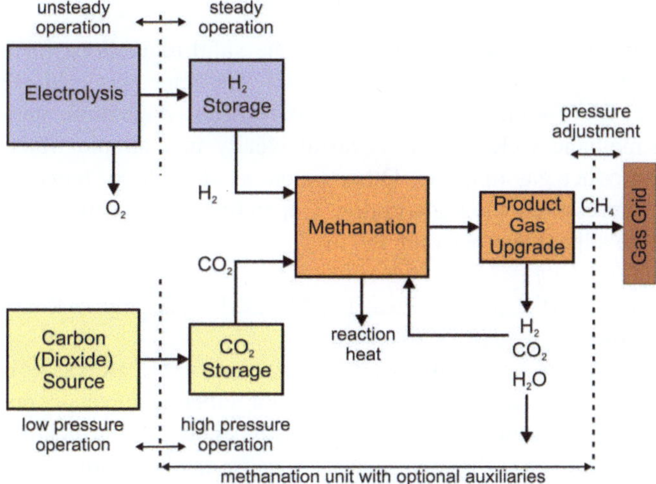

Fig. 4.5 Scheme of a chemical methanation unit within the Power-to-Gas concept

to hydrogen. Due to the high critical temperature of 31 °C, CO_2 can be lique-fied by compression. Whereas conventional methanation processes and catalysts have been developed for carbon oxide as feed gas, Power-to-Gas methanation uti-lizes carbon dioxide as educt. In terms of the catalysts it is known that both car-bon dioxide and carbon monoxide can be handled as feedstock. Potential carbon dioxide sources and their impact on the methanation process are discussed more detailed in the next chapter.

The educt gases—hydrogen and carbon dioxide—have to be compressed to the operational pressure of the methanation system. Electrolysis is already operated with elevated pressures depending on the utilized technology (see chapter electrol-ysis). In contrast, carbon dioxide sources are almost always at atmospheric pres-sure, and thus need compression in case of chemical methanation.

Typical conventional methanation plants (Fig. 4.2) are operated on an industrial scale and with a high annual availability. Methanation units as part of the Power-to-Gas concept may vary in plant size (several 100 kW to several 100 MW) and also in annual operation hours. Consequently, the methanation process and the reactor concepts have to be developed with regard to these boundary conditions: modular, easily up-scalable reactor designs, load flexible systems as well as pro-cess concepts capable of stand-by operations are favored. None of the conven-tional methanation process developments fulfill all of these requirements.

The improvement of the economic viability is one of the main objectives of methanation process developments for Power-to-Gas applications. The cost effec-tiveness of methanation can be positively influenced

- by reducing the efforts for the gas upgrade downstream of the reactor,
- by utilizing the released reaction heat within the Power-to-Gas process chain and outside of the system, respectively,
- by increasing the lifetime of the catalysts,
- and by achieving high annual operational hours.

The product gas upgrade aims to cope with the valid regulations for injection of substitute natural gas, and biogas, respectively, into the gas grid (for example DVGW G262, ÖVGW G31, see also Table 4.7). Multi-stage methanation reactors enable high methane yields and thus result ideally in a simple water condensa-tion unit as product gas upgrade. Other potential upgrade systems are based on membranes or pressure swing adsorption. Depending on the entry point to the gas grid, a pressure adjustment between the methanation unit and local grid pressure is required.

The utilization of the reaction heat is addressed separately in the chapter heat integration below.

Catalyst lifetime is both influenced by the content of catalyst poisons in the educt gases, and catalyst destruction by reactor inherent mechanisms (abrasion, hot spots). In order to minimize the efforts for the educt gas purification, extensive research is carried out for the identification of new catalytic substances which are less sensitive to typical catalyst poisons and which are more selective for carbon dioxide as feedstock (iC^4 2014).

Plant size, reactor design, set-up of the process chain and annual operating hours of a methanation unit within the Power-to-Gas concept substantially depend on the specific local conditions: available quantity and temporal profile of renewable power and thus hydrogen production, carbon dioxide source as well as size, pressure level and load flow of the natural gas grid. Therefore, each methanation unit of a Power-to-Gas process chain has to be tailored to the specific boundary conditions of a distinct application. This emphasizes the importance of the flexibility of both the methanation reactor systems and processes.

As already mentioned in the chapter "biological process routes", biological methanation promises to be advantageous compared to the chemical routes. Basically, the moderate operation conditions make the biological approach cost-effective. But, compared to chemical methanation, biological methanation still lacks maturity, and the expected advantages have to be proven in practice and on an industrial scale.

4.2.1 Process Educts: Hydrogen and Carbon Dioxide

Hydrogen is generated by the electrolysis unit upstream from the methanation. This production process secures a high pureness of the hydrogen, i.e. 99.99 %+ in case of PEMEC. The main impurity to be expected is oxygen. Therefore, in case of hydrogen, the feed gas composition is secondary compared to the temporal fluctuations of its production.

For the second feed gas, carbon dioxide, the composition with source inherent impurities is relevant, beside the capture costs itself. From a technical point of view, chemical methanation requires a minimum gas quality. In Table 4.3 the required gas quality is summarized in terms of primary and secondary components. Assuming a stoichiometric mixture of hydrogen and carbon dioxide with a molar ratio of 4:1 (see Eq. 4.3), a minimum gas quality for the carbon dioxide input can be estimated. This minimum quality requirement is also listed in Table 4.3.

In terms of the main components, the requirements are comparatively low. Since conventional methanation catalysts convert both carbon dioxide and carbon monoxide, basically both gases can be used as a carbon source. Considering a positive CO_2 balance of the Power-to-Gas system, the carbon source is either derived from biomass (biogas plant or biomass gasification) or from industrial CO_2 sources (fossil power plants, energy intensive industries). Another, quite extensive possibility is to utilize CO_2 from the atmosphere (ZSW 2014). According to the law of mass action, methane and steam concentrations in the educt gases should be minimized (see Eq. 4.3). Nitrogen is an inert gas in the methanation process. With regard to the plant size and the heat management of the system, the nitrogen content should be as low as possible. Particularly in terms of the product gas upgrade downstream from the methanation, nitrogen, but also oxygen, is challenging due to a lack of appropriate conditioning processes. Thus, air components should be

Table 4.3 Necessary gas quality for methanation (Müller-Syring et al. 2013; Bajohr 2014)

Component	Unit	Value for methanation input	Value for CO_2 stream
H_2	Vol.%	35–80	–
CO_2	Vol.%	0–30	0–100
CO	Vol.%	0–25	0–100
CH_4	Vol.%	0–10	0–50
N_2	Vol.%	<3	<15
O_2	Vol.%	n.s.	n.s.
H_2O	Vol.%	0–10	0–50
Particles	mg/scbm	<0.5	<2.5
Tar	mg/scbm	<0.1	<0.5
Na, K	mg/scbm	<1	<5
NH_3, HCN	mg/scbm	<0.8	<4
H_2S	mg/scbm	<0.4	<2
NO_x	mg/scbm	n.s.	n.s.
SO_x	mg/scbm	n.s.	n.s.
Halogens	mg/scbm	<0.06	<0.3

n.s.: not specified, *scbm:* standard cubicmeter (20 °C, 0.1 MPa)

minimized in the educt gases. Higher oxygen fractions may also negatively influence the catalyst activity and may promote undesired side reactions, but distinct concentration limits are still unknown.

The secondary components listed in Table 4.3 act predominantely as catalyst poisons, and should therefore be strictly limited. The difficulties to comply with the listed threshold values of the different secondary components depend on the origin of the CO_2 source, particularly tar, ammonia, particles and hydrogen sulfide are challenging. Limit values for NO_x and SO_x are still under investigation, and are not available at present. Obviously, the development of catalysts, which are tolerant for those secondary components, is important, because costly gas cleaning procedures can be avoided.

The supply of carbon dioxide is technically feasible, but is coupled with significant costs which have a considerable impact on the total costs of methanation. Basically, absorption, adsorption and membrane processes as well as carbonate looping are carbon capture options (Ausfelder and Bazzanella 2008; Scherer et al. 2012; Schneider et al. 2013). Amine based absorption processes are technically mature. The capture costs currently range between 25 and 60 €/t CO_2, the costs for an emission certificate within the EU trading system is presently about 5 €/t CO_2. Therefore, carbon capture is currently economically unattractive. In the case of biological methanation, untreated biogas can be used as a carbon source which has significant advantages for the cost structure. But the type of the carbon source has also to be adjusted to the size of the methanation unit. Large-scale units would need correspondently large-scale CO_2 sources which are not available as biogas.

4.2.2 Heat Integration

The aim of the heat integration is the coupling of the released heat of the methanation reaction with the required thermal energy for the CO_2 capture process. Thus, the economy of the system can be improved by energy savings for the CO_2 separation and by decreasing the cooling demand of the methanation reactor. The possibility of heat integration between the methanation and the carbon capture process has been simulated by Fraubaum and Haider (2014) with ASPEN. The considered process consists of a electrolyser and a downstream methanation. The rejected heat of the electrolyzer can only be utilized in district heating systems due to its temperature level in the range of 60–80 °C.

CO_2 capture and methanation can be combined by means of a steam turbine process. Since possible carbon sources origin from industrial processes, like fossil power plants, steel plants or cement mills, often steam power plants already exist, and therefore only adaptation of (but no new investments in) steam turbines are necessary. Precondition for the heat integration is the temporal decoupling of the electrolysis and the methanation. It can be achieved by a hydrogen storage tank (Fig. 4.5) or by a high temperature heat storage facility. Two ASPEN models have been elaborated, the first one is based on three adiabatic equilibrium reactors in series following the TREMP™ process, the second is based on an isothermal fluidized bed reactor (Comflux, see Table 4.1). As an example, a Power-to-Gas plant has been simulated with a connected duty of 100 MW_{el}. Calculated smaller duties show proportional behavior. The assumptions for the simulation are given in Table 4.4.

TREMP™ methanation consists in this simulation of three adiabatic fixed bed reactors. The gases are heated by the reaction heat in each reactor, and therefore have to be cooled down between each reactor step. The released heat can be used to produce superheated steam (65 bar, 400 °C).

Comflux methanation is operated in an isothermal fluidized bed reactor. Highly pressurized saturated steam (120 bar, 324.6 °C) can be produced with the released reaction heat. Table 4.5 summarizes the heating and cooling flows of a 100 MW_{el} Power-to-Gas system for both methanation types. In both cases, about 1 kg/s of CH_4 (which corresponds to 5,000 m^3/h SNG under standard conditions) is produced with the simplified assumption of total educt conversion.

In both cases, the produced steam has a significantly higher energy level than required for the CO_2 desorption (2 bar, 120.3 °C). Therefore, the produced steam can be expanded in a condensing turbine. The resulting turbine power is listed in Table 4.6 for different connected loads of the electrolysis. The expanded steam is fed to the CO_2 stripper and is condensed. In a conservative estimate, the heat demand for CO_2 stripping is set to 3.5 GJ/t CO_2 with an assumed separation rate of 90 % (Table 4.4). Obviously, the released heat supersedes the heat demand of CO_2 desorption by far, even the heat content of the expanded steam is higher than required for the stripping process. Thus, still heat losses occur by condensation cooling in the steam circuit. Cooling is also required for the conditioning of the produced SNG (condensation of the water content).

Table 4.4 Assumptions for the simulation of heat integration

Electrolysis		
Connected duty	MW	100
System efficiency	%	70
Released heat	MW	1/3 of connected load
Temperature of H_2	°C	70
CO_2 capture		
Separation rate	%	90
Heat demand	GJ/kg CO_2 capt.	3.5
Demand of electricity	GJ/kg CO_2 capt.	0.1
Steam pressure for regeneration	bar	2
Pressure of CO_2	bar	2
Pressure loss	bar	0
Isentropic efficiency (compressor)	–	0.72

Methanation			
Methanation process		TREMP™	Comflux
Reactor entrance temperature	°C	300	400
Reactor pressure	bar	27	27
H_2/CO_2 ratio	mol/mol	4	4
Reflux rate	weight-%	69	0

Steam circuit			
Methanation process		TREMP™	Comflux
Vapor pressure	bar	65	120
Temperature saturated steam	°C	400	324.6
Isentropic efficiency (turbine)	–	0.87	0.87
Turbine exit pressure	bar	2	2
Pinch point	°C	10	10

Table 4.5 Simulated heating and cooling flows of a 100 MW Power-to-Gas system

Methanation process	Released heat of electrolysis [MW]	Required heat for CO_2 desorption [MW]	Totally released reaction heat [MW]	Preheating of educt gases [MW]	Heat losses by cooling	
					Steam circuit [MW]	Conditioning of SNG [MW]
TREMP™	33.4	9.44	14.5	1.36	1.58	1.53
Comflux	33.4	9.44	13.9	2.04	1.13	2.52

Table 4.6 Resulting turbine power for different connected loads of the electrolysis

Connected load to electrolyser [MW_{el}]	Resulting turbine power for TREMP™ methanation [MW]	Resulting turbine power for Comflux methanation [MW]
100	3.53	3.43
75	2.66	2.52
50	1.77	1.72
28	1	1

A connected duty of the 100 MW$_{el}$ electrolyser is comparatively high, result-ing in a turbine power of 3.5 MW. It is possible to operate turbines with a high efficiency of 80–90 % with ratings down to 1 MW which correspond to approx-imately 30 MW electrolyser power. Smaller steam turbines lack efficiency. Therefore, the heat integration of a steam turbine process in Power-to-Gas systems smaller than 30 MW has to be considered critical.

4.2.3 Development Trends and Current Research Activities

The development trends in methanation as part of Power-to-Gas can be summa-rized as follows:

- Improvement of catalysts regarding their durability under poisoning conditions and towards better selectivity also for carbon dioxide as feedstock. Extensive research is conducted at the moment by several researcher groups (iC^4 2014; Lehner and Steinmüller 2013). It is also necessary to obtain an improved insight into the reaction mechanisms, and to derive kinetic formulations for the metha-nation reaction.
- Investigation of the methanation process regarding the ability of intermittent and dynamic operation, its connection with other elements of the Power-to-Gas system, as well as possibilities of heat integration. The basic target is an increase of the efficiency, the durability and the flexibility of the methanation process (Lehner and Steinmüller 2013; Energie-Agentur 2012).
- Development of innovative reactor concepts which cope with the specific requirements of the methanation process, i.e. the highly exothermic reaction, a modular design for easy up-scale, a potentially high load flexibility, low standby energy requirements as well as short deployment and cold start times. In this context, two reactor concepts are promising: the three-phase methanation sys-tems (Bajohr et al. 2011; Graf 2013), and monolithic catalyst carriers imple-mented in staged reactors (Bajohr et al. 2011; Lehner and Steinmüller 2013; Graf 2013). Particularly, monolithic catalysts enable a modular, easily up-scal-able design, reduced mechanical stress to the catalyst due to the static catalyst bed, reduced pressure loss, and, owing to the staged reactor design, a uniform temperature profile. Both monolithic carriers made from metal or ceramics are currently under investigation. Typically, such basic investigations are carried out in laboratory scale plants, as shown in Fig. 4.6 in which 3 reactors are operated in series with gas cooling, gas sampling and recirculation between each reactor.
- In case of the biological methanation, the investigation of the long-term stabil-ity as well as a fully operational control of the systems has to be investigated. Furthermore, an up-scale to the MW scale is necessary to give this technology a wider range of application. Operational experience in the field of practice has to be obtained by implementing and operating pilot plants.
- Decrease of product gas upgrade complexity downstream from the methana-tion unit in order to achieve product gas qualities which comply with the speci-fications for the natural gas grid (see Table 4.7). It can be achieved either by

Fig. 4.6 Methanation
laboratory plant at
Montanuniversität Leoben
during assembly

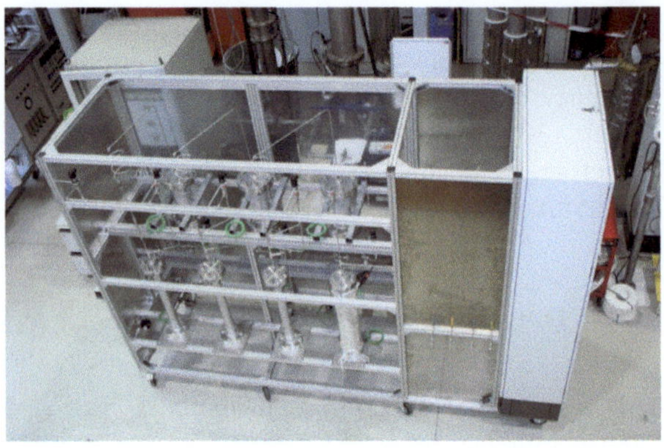

Table 4.7 Specification of the SNG composition according to ÖVGW guideline G31

Parameter	Unit	Permitted range for SNG
Wobbe index	kWh/m^3	13.3–15.7
Heating value	kWh/m^3	10.7–12.8
Oxygen (O_2)	Mole %	$\leq$0.5
Carbon dioxide (CO_2)	Mole %	$\leq$2.0
Nitrogen (N_2)	Mole %	$\leq$5
Hydrogen (H_2)	Mole %	$\leq$4
Total sulfur	mg S/m^3	$\leq$10
Mercaptan sulfur	mg/m^3	$\leq$6
Hydrogen sulfide	mg/m^3	$\leq$5
Carbon oxide sulfide (COS)	mg/m^3	$\leq$5
Halogen compounds	mg/m^3	0
Ammonia (NH_3)		Technically pure
Solid and liquid components		Technically pure

improved reactor or process concepts, or by specifically tailored upgrade processes. Membranes are currently under investigation, for example, and show promising results (Makaruk 2011).

- Determination of the potential and the costs of different CO_2 sources considering the competition with the utilization in other sectors, as well as alternative sources from energy extensive industries. Investigation of the link between chemical and biological methanation with biogas plants.
- Erection and test of pilot and demonstration plants in order to gain practical and long-term experience in different fields of application. A list of currently operated or planned projects in Germany can be found at DVEW[3] projects in other European countries in Grond (2013). Most of the Power-to-Gas demonstration projects do not include the methanation step.

---

[3] http://www.dvgw-innovation.de/presse/power-to-gas-landkarte/. Accessed 30 April 2014.

4.2.4 Actual Costs and Future Cost Development Potentials

Generally, due to the early stage of development, little authoritative data is available for an assessment of the costs. Furthermore, the determination of investment costs is strongly influenced by the specific plant layout and the variety of operational conditions. Operational expenditures are determined by the costs of CO_2 capture, the possibilities for utilization of the product methane and the by-products (i.e. released heat), the annual operational hours, plant size (economy of scale) and existing infrastructure on site (personnel, safety infrastructures etc.). The current uncertainties in cost assessment underline the necessity of demonstration plant operation to gain a better insight also into the cost structures.

Based on the current state of the art of electrolysis (alkaline electrolyser) and methanation (fixed bed), a study has evaluated the cost structure for a connected duty of 48 MW_{el} (Kinger 2012). The total investment costs for electrolysis, methanation and auxiliaries comprise 1,000 €/kW_{el}, in which 86.3 % are allocated to the electrolysis. Consequently, the costs of methanation amount to 140 €/kW_{el}. The investment costs of a 5–10 MW_{el} demonstration plant amounts to 2,000 €/kW_{el} (Sterner 2009), and are assumed to reduce to 1,000 €/kW_{el} for greater plant sizes. This investment covers the electrolyser, the methanation, the gas compression, power electronics, piping, civil works and control systems. Assuming the same cost structure as found in Kinger (2012), the methanation investment costs can be calculated in the range of 135–275 €/kW_{el}. The investment costs of methanation plants smaller than 10 MW_{th} are depicted in Fig. 4.7, both for chemical and for biological methanation. The data is taken from Grond et al. (2013). For the future, the authors assume a cost degradation for plants smaller than 10 MW_{th} due to standardization of smaller plants which results in a cost for chemical methanation of 300–500 €/kW_{CH4}. It should be noted that the conversion between €/kW_{el} and €/kW_{CH4} can be achieved by multiplication with the efficiencies of electrolysis and methanation (i.e. 70 % for electrolysis and 80 % for methanation).

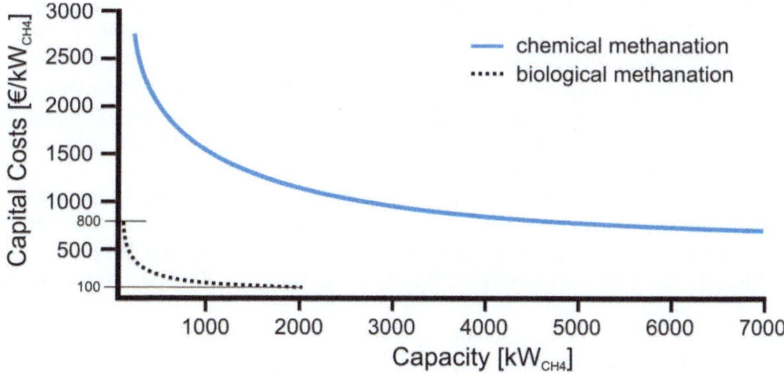

Fig. 4.7 Investment costs of chemical and biological methanation, data taken from Grond et al. (2013)

Thus, 300–500 €/kW$_{CH4}$ are equal to 160–280 €/kW$_{el}$, and therefore in the same range as in Kinger (2012) and Sterner (2009).

The comparatively low costs of biological methanation (Fig. 4.7) can be justified with the moderate operation conditions (atmospheric pressure, temperatures <70°C), resulting mainly in cheaper materials of construction. Also, no catalyst is needed, and a purification of the feed gases is not necessary or is simpler. On the other hand, there is still a lack of operational experience for biological methanation, and plant size is still limited to the lower MW range.

The annual operation and maintenance costs are assumed to be 10 % of the capital costs (Grond et al. 2013) what appears to be somewhat high. Usually, operation and maintenance costs amount to between 3 and 7 % of the investment costs.

The above introduced cost structures may be changed significantly in the future, for example by improvements in electrolyser systems or efficiencies. In any case, the future cost development depends on the progress in methanation reactor design, process conditions (pressure, temperature), catalyst's lifetime and selectivity, and particularly in utilizing the possibilities of integration and symbiosis with other industrial plants (heat integration, utilization of produced by-products like oxygen, common infrastructure).

References

Elvers B et al (ed) (1989) Ullmann's encyclopedia of industrial chemistry, vol A12, 5th edn. VCH Weinheim, New York

Mills GA, Steffgen FW (1974) Catalytic methanation. Catal Rev 8(1):159–210

Kaltenmaier K (1988) Untersuchungen zur Kinetik der Methanisierung von CO_2-reichen Gasen bei höheren Drücken. Dissertation, Universität Karlsruhe

Wang W, Wang S, Ma X et al (2011) Recent advances of catalytic hydrogenation of carbon dioxide. Chem Soc Rev 40(7):3703–3727

Weatherbee GD, Bartholomew CH (1982) Hydrogenation of CO_2 on group VIII metals. II. Kinetics and mechanism of CO_2 hydrogenation on nickel. J Catal 77(2):460–472

Bartholomew CH (2001) Mechanism of catalyst deactivation. Appl Catal A 212:17–60

Kopyscinski J, Schildhauer TJ, Biollaz SM (2010) Production of synthetic natural gas (SNG) from coal and dry biomass—a technology review from 1950 to 2009. Fuel 89(8):1763–1783

US Department of Energy (2014) Practical experience gained during the first twenty years of operation of the great plains gasification plant and implications for future projects. http://www.netl.doe.gov/research/coal/energy-systems/gasification/gasifipedia/great-plains. Accessed 26 Mar 2014

Bajohr S, Götz M, Graf F, Ortloff F (2011) Speicherung von regenerativ erzeugter elektrischer Energie in der Erdgasinfrastruktur. gwf-Erdgas:200–210

Haldor Topsøe A/S (ed.) (2009) From solid fuels to substitute natural gas (SNG) using TREMP™. http://www.topsoe.com/business_areas/gasification_based.aspx. Accessed 26 Mar 2014

GoBiGas (2014) http://gobigas.goteborgenergi.se/En/Start Accessed 29 Mar 2014

Seemann MC, Biollaz SMA, Aichernig C, Rauch R, Hofbauer H, Koch R (2004) Methanation of biosyngas in a bench scale reactor using a slip stream of the FICFB gasifier in Güssing. In: proceedings of the 2nd world conference and technology exhibition: biomass for energy, industry and climate protection, Rome

Biollaz SMA, Schildhauer TJ, Ulrich D, Tremmel H, Rauch R, Koch M. (2009) Status report of the demonstration of BioSNG production on a 1 MW SNG scale in Güssing. In: Proceedings of the 17th European biomass conference and exhibition, Hamburg

Chem Systems Inc. (1979) Liquid phase methanation/shift pilot plant operation and laboratory support work. Final report, 1 July 1976–30 Nov 1978. Prepared for US Dept of Energy, Contract No. Ex-75-C-01-2036

Karakashev D, Batstone DJ, Angelidaki I (2005) Influence of environmental conditions on methanogenic compositions in anaerobic biogas reactors. Appl Environ Microbiol 71(1):331–338

Luo G, Johansson S, Boe K, Xie L, Zhou Q, Angelidaki I (2012) Simultaneous hydrogen utilization and in situ biogas upgrading in an anaerobic reactor. Biotechnol Bioeng 109(4):1088–1094

Luo G, Angelidaki I (2012) Integrated biogas upgrading and hydrogen utilization in an anaerobic reactor containing enriched hydrogenotrophic methanogenic culture. Biotechnol Bioeng 109(11):2729–2736

Schaub G, Iglesias Gonzales M, Eilers H (2014) Chemische Reaktoren als Elemente eines flexiblen Energiesystems? Presentation at Fachausschuss Energieverfahrenstechnik, Karlsruhe 18 Feb 2014

iC^4—Integrated Carbon Capture, Conversion & Cycling (2014) http://www.ic4.tum.de/. Accessed 20 Apr 2014

ZSW (2014) Verbundprojekt "Power-to-Gas": Errichtung und Betrieb einer Forschungsanlage zur Speicherung von erneuerbaren Strom als erneuerbares Methan im 250 kW$_{el}$-Maßstab. http://www.zsw-bw.de/fileadmin/editor/doc/20111019_Power-to-Gas_Projektinfo_01.pdf. Accessed 21 Apr 2014

Müller-Syring G et al (2013) Entwicklung von modularen Konzepten zur Erzeugung, Speicherung und Einspeisung von Wasserstoff und Methan ins Erdgasnetz, DVGW Bericht zu Fördezeichen G 1-07-10:119

Bajohr S (2010) Methanisierung—technische Ansätze und deren Bewertung. International biomass conference, Leipzig

Ausfelder F, Bazzanella A (2008) Verwertung und Speicherung von CO_2. Dechema, Frankfurt am Main

Scherer V, Stolten D, Franz J, Riensche E (2012) CCS-Abscheidetechniken: Stand der Technik und Entwicklungen. Chem Ing Tech 84(7):1026–1040

Schneider G, Schneider R, Hohe S (2013) Technical challenges and cost reduction potential for post-combustion carbon capture. In: Proceedings of power gen Europe, Vienna

Fraubaum M, Haider M (2014) Analyse des Wärmemanagements des Gesamtprozesses. In: Steinmüller H et al (2014) Power-to-Gas—eine Systemanalyse. Endbericht, Wien

Lehner M, Steinmüller H (2013) Power-to-Gas Eine Option zur chemischen Speicherung und zum Transport von erneuerbaren Energien. In: Biedermann H, Vorbach S, Posch W (eds) Ressourceneffizienz Konzepte, Anwendungen und Best-Practice Beispiele. Rainer Hampp Verlag, München und Mering

Deutsche Energie-Agentur (ed) (2012) Eckpunkte einer roadmap power to gas. http://www.powertogas.info/roadmaps.html. Accessed 30 Apr 2014

Makaruk A (2011) Numerical modeling, optimization and design of membrane gas permeation systems for the upgrading of renewable gaseous fuels. Dissertation, Technische Universität Wien

Grond L, Schulze P, Holstein S (2013) Systems analyses power to gas: deliverable 1: technology review. DNV KEMA Energy & Sustainability, Groningen

Graf F (2013) Forschungsaktivitäten zur Methanisierung am Engler-Bunte-Institut des Karlsruher Instituts für Technologie (KIT). Presentation at Innovationsforum PGP, Leipzig 24 Apr 2013

Kinger G (2012) Green energy conversion and storage (Geco). Endbericht for FFG project 829943, Wien

Sterner, M (2009) Bioenergy and renewable power methane in integrated 100 % renewable energy systems. Dissertation, Universität Kassel

Chapter 5
Business Models

This chapter (Tichler 2014)[1] focuses on the economic dimension of the Power-to-Gas technology respectively of the Power-to-Gas system. The economic dimension can be defined as quite complex due to the fact that Power-to-Gas offers significant benefits to the whole energy system. As a consequence, an economic analysis of Power-to-Gas requires not only business analysis but also comprehensive macroeconomic and systemic analysis. Only the combination of those two approaches offers an all-embracing method for statements on economic characteristics of Power-to-Gas.

In addition, technologic developments and characteristics determine the economic characteristics but also vice versa. This is quite more important in cases where the shape of the focused system offers significant space for additional improvements—like in the case of Power-to-Gas. This gives additional necessity to economic analysis.

As a consequence, the following chapter starts with systemic and macroeconomic impacts with special focus on the context of storage systems and new possibilities for energy transport. That is followed by an analysis of several process chains of Power-to-Gas and leads to the analysis of specific business models.

5.1 Macroeconomic and Systemic Impacts

Owing to the defined national and international objectives, the proportion of renewable energy sources for providing electricity will rise in the next decades even in the event of an increase in energy consumption in countries of the European Union

---

[1] Significant parts of this chapter was funded and co-funded by the Austrian ministry of economics, by Österreichs Energie and by FGW. The report (Steinmüller et al. 2014) is only available in German language and can be ordered from Energieinstitut an der Johannes Kepler Universität Linz.

© The Author(s) 2014
M. Lehner et al., *Power-to-Gas: Technology and Business Models*,
SpringerBriefs in Energy, DOI 10.1007/978-3-319-03995-4_5

as well as in many other regions. Focusing on the sustainability of the energy system and reducing dependence on the import of fossil fuels, wind turbines and photovoltaic power plants are increasingly being built. The environmental and energy objectives of increased integration of renewable energy into the present energy sector—and in particular in the production of electricity—generates new challenges to economies, especially in providing a high level of supply security. This challenge results in particular from strong fluctuations in the supply of renewable energy from volatile sources as wind and solar power. In consequence, as the share of these energies increases, a balance between power surpluses during periods of high levels of generation and periods of shortfalls is necessary (Tichler 2011a).

Energy storage systems will play a key role providing an integration of renewable energy sources with volatile production structures in addition to optimized power management. The storage of energy, especially of electrical energy, and the availability of the correct amounts of energy at the proper time period present a major challenge. The Power-to-Gas technology will be an important part in future storage portfolios, because long term storage as well as shifts in capacity between energy networks can be realized, which offer new possibilities in energy transmission.

Furthermore, Power-to-Gas systems can solve additional energy and environmental policy challenges presented by the enlargement of the percentage of alternative fuels in the mobility or heating sector. In the long term Power-to-Gas technology can provide a significant shifting in the use of produced energy out of volatile sources from conventional power usage to dependable usage in the form of methane and hydrogen. The spread reconversion of gas for electricity supply cannot be excluded in this context, but should not be preferred in terms of energetic reasons, because of the related loss of efficiency and increase in costs.

These additional resulting forms of electrical energy usage by converting hydrogen and methane via Power-to-Gas technologies correspond to general forecasts of the future long-term substitution within primary energy carrier. The energy carrying gas and its different characteristics is represented in many forecasts for future power supply as the essential transitional technology or immanent energy carrier on the road to the "hydrogen economy" (Tichler 2011a). Furthermore, loss of importance of liquid energy sources as oil and solid fuels like wood, coal and uranium are prognosticated by Hefner (2007) in the global context (Tichler 2011b). As a consequence the "Age of Energy Gases" is starting in this decade as shown by Hefner (2007) with respect to the future development of the global composition of energy suppliers.

The central question for the implementation of new technology in general is, whether it is profitable already. That issue is analyzed with all aspects in the following chapter. Therefore a financial viability is compulsorily relevant for each technology. In case of existing economic relevance, further development and implementation of a technology can be of prime importance. The realization of market penetration is forced by economical reliability. The Power-to-Gas system operates in this field.

It is illustrated that the Power-to-Gas system generates a number of parameters, which offer a superior benefit for the energy system. The system is represented for the Austrian economy and offers benefits, for example, in the form of increased supply security or contributions to emissions reduction.

A fundamental direct economic improvement of the system can also be generated by knowledge and technology development in addition to the improvement and optimization of the energy system, technological substitution by domestic products at home as well as technology export with innovative products must be pursued.

5.1.1 Potential Solutions of Energy Storage: Power-to-Gas in the Context of Alternative Storage Systems

In 2014 many energy markets are confronted with high growth rates of wind and solar power and therefore a growing problem of time-dependent production fluctuations. As a result, occurring regional surpluses have to be transmitted through an appropriate grid infrastructure. The electricity generation from wind as well as solar power present pronounced fluctuations throughout the course of a day or year and with relatively limited predictability. Additionally, the usage of wind power is concentrated regionally, such as in Mecklenburg Vorpommern (Germany) or in Burgenland (Austria). The development of the power girds required for this purpose is combined with significant landscape interferences. Due to this fact, socio-democratic problems can be expected. Although, as the expansion of the transmission grids represents—even if problematic—an opportunity to balance production fluctuations of renewable energy sources, with compensation via the grid, this solution is neither economically useful nor technologically possible.

Due to a significant rise of the power consumption in the past (except for the years of the financial crisis in Europe from 2008 to 2010) in different sectors of the European economy, power grids are reaching their limits in a considerable number of hours. As a result the risk of recent blackouts with significant consequences to population and economy is continuously rising. On the one hand, adequate investments in expansion and maintenance of power grids and on the other hand long term concepts for compensation of large load balancing are a basis to provide a high level of supply security. The importance of the involvement of additional storage technologies becomes increasingly obvious.

In some countries of Europe, for example in Germany or Austria, the production of power out of renewable energy sources reached an extent, that in specific times significant surpluses of power are present from huge amounts of production out of wind and solar energy. The compilation of the European power plants with a relatively high share of nuclear, water and wind power plants, could hardly reduce their production level in case of an oversupply of electricity, which represent a rising challenge to the energy system. It will be necessary that in times of high power output from volatile production sites as wind power plants to take measures for grid stabilization in the form of energy storage systems. As an example, the situation in Burgenland is described here. In Burgenland power input out of renewable energy sources can temporarily significantly exceed the power consumption and on the other hand the input only covers about 20 % of the consumption in certain periods.

Table 5.1 Potential of wind power expansion in Eastern Austria according to the energy storage amount and the possible resulting hydrogen-capacity for the mobility in Austria

Parameter	Unit	Value
Approximate wind power expansion in East-Austria to 2030	MW	3,000
Full load hours	h/a	2,000
Electrical production wind power expansion in East-Austria	MWh/a	6,000,000
Proportion fluctuation and available electrical production according to the expansion (assumption)	%	40
H_2-storage potential	MWh/a	2,400,000
Conversion efficiency of the electrolysis	%	60
H_2-energy Supply	MWh/a	1,440,000
H_2-energy density	kWh/kg	33.30
H_2-supply	t/a	43,243
H_2-consumption of a typical vehicle	kg/100 km	1.2
Average annual kilometers per car	km/a	12,000
With H_2-Production compatible cars	piece/a	300,300

Source own calculation, (Amt der NÖ Landesregierung 2011), (www.energieburgenland.at/oekoe nergie/windkraft/. Accessed 30 May 2014) NÖ Energiefahrplan, Energie Burgenland, Opel

Based on the analysis of the expansion plans and concepts for wind power plants in the area of Lower Austria and Burgenland, an energy storage amount of 2.4 TWh$_{el}$ can be expected. This assumes that all plans are completely implemented so that 40 % of the produced energy amount is provided by volatile production from additional wind power plants (without taking account of any existing plants). Under these assumptions, a hydrogen production of about 43,000 tons per annum out of Power-to-Gas-systems is calculated in accordance to Table 5.1, which can supply 300,000 hydrogen powered cars.

An alternative concept instead of expanding grid architectures is represented by energy storage with transportation via natural gas networks and reconversion in periods of higher demand. Power-to-Gas can store excess power by converting the electrical energy into hydrogen or methane used as an energy carrier. The gases can be fed into the natural gas grid and used in all areas where gas distribution systems are present. Pumped-storage plants as well as adiabatic compressed air reservoir storage are used as a benchmark for the storage of electrical energy. These storage technologies can provide long term energy storage with limitations as to the reconversion into electrical power. Pumped-storage plants are currently the almost exclusively present storage technology, including their topographical ligation and the resulting need for transportations through the electrical grid to the storage units. High volume electrochemical storages such as batteries are in discussion and partly also in the testing phase as well as pumped-storage plants.

Chemical Storage of electrical energy in form of hydrogen or synthetic methane generates systemic storage benefits:

- On the one hand storage can be provided on-site, for example directly to the wind park, so investments in grid architecture can be substituted.

- On the other hand big storage sites suitable for methane are already present in Europe, so existing infrastructure can be used by coupling the electrical and gas grid. The usage of such available storage enables a further expansion of the European energy storage system. The European overall economy has the capability, to use already existing large-scale natural gas storage facilities.
- Furthermore electrical grid expansion can be avoided when stored energy is transported away through the gas distribution system.

Generally electrical energy storage units have contributed to ensuring a secure and low-cost energy supply over the last decades. This applies to storage units at the local level and above and also for central mass storage systems, which play an important role for the general electrical supply: a time decoupling between electrical power generation and consumption can be reached via storage systems, which makes it possible to run capital-intensive base load plants also in low-load periods. On the other hand storage can provide network and ancillary services through a high level of controllability. In this way storage represents an important contribution to ensure a stable operation of the electrical power grid. Economically these tasks are mainly implemented in pumped-storage plants and have been for many decades.

Through further expansion of renewable energy production, overall structural and operational requirements to the production system are increasing. Especially the integration of fluctuating and limited scope for forecasting power production out of wind and solar plants requires a broad adaptation of the energy system, in particular the expansion on local, regional, national and transnational levels. Against this background, significant efforts have already been undertaken until recently, both at national and international level, to expand existing pumped-storage plants and introduce new storage technologies to the market.

Nowadays it is already reflected, that the technical and organizational structures of the power supply system restrict change to efficiently integrate the fluctuation rate of the renewable energies. To provide long-term sustainable as well as secure and cost-effective electricity supply, the adaptation of the system is required so that renewable energies can coordinate their production with demand and the available network and storage capacities in addition to the creation of new energy storage. This can be realized by the integration of chemical energy storage or Power-to-Gas systems. Relating to the creation of Power-to-Gas systems beside production units with volatile production, electrical energy can be saved in optimal periods—such as night hours—transported through the power lines, or supplied to gas distribution systems in the form of hydrogen or methane.

Despite the widely recognized long-term needs of additional storage capacities parts of the population as well as nature conservation and environmental organizations critically evaluate the new construction of pumped-storage plants due to the selective massive impact on the landscape (Cohen et al. 2014). In many cases decentralized storage technologies are required and perform better due to reduced environmental impact from the point of view of local residents near to pumped-storage projects. Further storage technologies must be introduced to the market in addition to pumped-storage projects because the locations and the storage capacities of additional pumped-storage are limited (Tichler 2011). As an example, the Austrian Alpine region which is topographically well suited for pumped-storage is explained here. The "Deutsche Energieagentur" quantifies for example the potential of additional pumped-storage in the Austrian Alpine region in their simulation calculation to 2 GW_{el} (dena—Deutsche Energie-Agentur-GmbH 2010). It is not known whether projects which have already been planned at the end of the year 2010 are included. "Österreichs Energie" specifies concrete pumped-storage projects with a planned installed capacity of 2.9 GW_{el} (Oesterreichs Energie 2010). In total, pumped-storage plants with an installed electrical performance of about 3.8 GW_{el} exist in Austria at the end of the year 2010 (Tichler et al. 2011). A complete coverage of the Austrian power requirement through renewable energy sources would mean a massive expansion of production capacities out of wind and photovoltaic energy. That in turn, would require storage systems which exceeds the existing capacity of pump storage by a factor of hundred (TU Wien 2011).

A further crucial advantage of pump storage plants is, in addition to the mature technology, the high efficiency compared to Power-to-Gas systems. Serious disadvantages are the high dependence on location due to the required difference between upper and lower reservoirs and the already mentioned acceptance issues raised by the local population in the case of such planned new constructions. Furthermore, in contrast to Power-to-Gas, the use as week or month long storage or as a seasonal redistribution system of the production quantities are not representable because the storage capacities compared to long-term expectable weekly or monthly power surpluses out of wind and photovoltaic plants are too low. As consequence, pump storage is mostly used for hours and days to balance the production and consumption fluctuations between day and night or working days and weekends and holidays as well as provide for regulation performance. This gives Power-to-Gas systems a significant potential due to their operational capability.

In contrast to pump storage, further storage technologies such as compressed air reservoir, gyrating mass, super capacitors, superconducting coils, and large-scale battery reservoirs, are mostly on the cusp of commercial introduction or in the development and demonstration stage. Compressed air reservoirs are used comparable to the power range and performance of pump storage. But the efficiency is comparatively small. A further development is presented by Adiabatic Compressed air reservoirs, which should reach efficiencies up to 70 %. The concept is still in the development phase. For the system integration of renewable energies in principle small decentralized storage can be connected to the distribution network as battery storage in addition to central mass storage. These storage systems are certainly less suitable to storage cycles in the week or month range. The energy storage of Power-to-Gas systems in chemical form enables the volatile incidental power production as a result of decentralized plants to be saved next to the renewable production plant, before feed-in into the gas distribution system takes place. These generate the possibility of the reduction of future stranded investments in fossil power plants. The feeding-in of electrical power produced by photovoltaic-modules at peak times, rapidly reduces the margins or the rentability of existing "conventional" power plants and it responds, amongst other things to quiescence of (gas-) power plants in consequence of this development (In addition to other parameters). The chemical energy storage could reduce the peak-supply of renewables and enable higher utilization of the existing power plants in this way.

5.1.2 New Possibilities for Energy Transport Due to Power-to-Gas Plants

A further central aspect of electrical energy storage in chemical form by the use of Power-to-Gas systems is the possible shifting of the energy transport from the electrical grid to the gas distribution system. This is relevant for both the production of hydrogen as addition to the natural gas network as well as for synthetic methane.

Power-to-Gas systems can also reduce a larger socio-demographic problem of the energy system by facilitating the intelligent site-placement of plants. The realization of large-scale high-voltage power lines through Europe or Central Europe that are faced with great resistance in the population could be to some extent avoided. The current surpluses increasingly occurring out of wind power have to be transported either directly to the buyers or to conventional electricity storage such as pumped storage. Therefore in the future huge investments in the expansion of the European power grid are expected. This expansion of power grids, as for the transport of electrical energy from the North Sea from North Africa or from the storage areas in Scandinavia, will be associated with significant changes to its topography, thereby providing a significant socio-demographic problem: a massive resistance in the population to new power lines. The acceptance of the public to major infrastructure projects that cause significant interventions in the landscape is currently hard to engender. These large power lines that will be connected most likely with metastases due to the relatively high degree of urban sprawl in Central Europe can be replaced by a shift of the energy transport from the power supply into the gas network, so that the transport does not have to be made via the power grid. This can reduce social tensions from necessary electricity infrastructure projects because the existing gas network can still absorb very large additional capacity without expanding. This is the well-known NIMBY ("not in my backyard") problem. In addition the realization of additional pump storage power stations is also associated with high acceptance problems in the society. In general, although the development of renewable energy sources from the general population is advocated, however, the necessary implementation of major infrastructure projects is confronted with local resistance.

With regard to the acceptance of Power-to-Gas installations there is as yet no evidence. In general, improving the acceptance of new technologies and energy infrastructure projects in society is essential, a quantitative assessment of new technologies, however, is very challenging. The level of acceptance of the Power-to-Gas technology cannot yet be dispensed with entirely due to the current research and development stage. In the context of market penetration, however, is the social acceptance of hydrogen technologies of high importance—i.e. in those application areas where people come into contact with the energy source in their environment and in their everyday use (e.g. mobility). For this purpose recent studies on the acceptance of hydrogen mobility in the Germany do exist. The first approximate conclusion of acceptance in society that has to be stated is that Power-to-Gas (H_2 and CH_4) has advantages over the existing alternative storage technologies as well as over network expansion. This represents a non-negligible component in the future prospects of the technology.

The partial shift of the energy transport from the electrical to the gas network would no longer make the required large-scale expansion of the electricity network necessary. The expansion is necessary due to energy policy developments, in particular the German energy transition. By converting and storing the energy in the form of hydrogen or methane, a form of energy is produced, which has a high energy density, and the generated energy can be transported in the already existing gas networks to the consumption centers.

The intensity of the necessary but also problematic electricity transmission network in Central Europe can be reduced. Additional fluctuations in the power grid increase the risk of large-scale power outages, with serious consequences for the population and the economy if no relevant compensation measures are taken. Various studies [for example (Brauner 2003; Reichl et al. 2006)] quantify the extent of cascading effects and the costs of large-scale power outages.

Thus, the threat to the implementation of alternative necessary reinforcement of the grid due to the already existing large public opposition to large power lines and train paths, due to the substantial interference with the landscape and the settlement areas through a wide application of the Power-to-Gas technology are reduced. The transfer of energy transport from the electrical to the gas network is summarized here in two dimensions:

1. The much higher energy density in the natural gas grid and the structure of the current natural gas network require no additional supply of significant capacity and no large-scale expansion of pipeline networks, so that investment in infrastructure can be reduced.
2. Also, an extension of the gas network would have in relation to the expansion of the electricity network a much lower topographical impact in which the acceptance of the population will increase and the land costs can be reduced. Thus, the construction of additional potential gas pipelines requires much less intervention in the landscape or settlements than power lines require with the identical amount of energy transported. Figure 5.1 illustrates this case. The space requirement of natural gas pipelines in comparison to power lines at equal energy transport volume.

This advantage leads to a large advantage in future energy policy and in structuring and designing the energy infrastructure. Here, the costs for expanding the electricity grid infrastructure and the implementation of a single broad power line are considered. Whereas nowadays the costs of the Power-to-Gas technology are quite

Fig. 5.1 Comparison of
floor space required for gas
pipelines and for power lines
by transporting the same
energy capacity. *Source*
(Energieinstitut an der
Johannes Kepler Universität
Linz GmbH 2012). *Note*
"Stromleitungen" = power
lines; "Unsichtbare
Gaspipeline" = unseeable
gas pipelines

high, the costs of Power-to-Gas can achieve market competitiveness in the long run
by including those additional costs into the whole system costs (Tichler et al. 2011).

Some studies also analyze that distribution from renewable energy sources due
to the increasing electricity production must be expanded, wind power in particular on-shore photovoltaic so as not to cause a collapse in the supply reliability of
the power system:

> The distribution must be extended in the course of the development of renewable energy
> [...]. Especially in areas with growing interests in solar and wind energy, where it often
> comes to recoveries from distribution to transmission network, there is a great need for
> adjustment (SRU 2010).

The problems arising in the local distribution networks due to the increasing distributed power generation are seen mainly in three areas:

1. Line overloads: Due to a faster aging from the congestion of NSP Cable
   Additions, very elaborate parallel cables are required.
2. Transformer overloads: Here is (albeit with increased occurrence) the identical
   problem as for line overloads, and additional transformers required.
3. Violations of the voltage band: In the distribution networks with a very frequent
   occurrence of this problem is expected, increasing risk of equipment damage
   exists. A possible but not yet sufficiently researched approach would be a reac-
   tive power control.

Schmiesing (2010) states that the transmission capacity of lines and transformers
is not the main problem but the primary problem is in fact the voltage stability
seen by the customer. Igel et al. (2010) state to solve the problems in distribution,
that when it exceeds the maximum line voltage of the network operators, in addition to the classical solutions, can also use an active voltage control and the use of

electric energy storage makes sense as a solution option (Tichler et al. 2011; Igel et al. 2010).

Thus, the Power-to-Gas technology through the allowed alternative energy transport capacity of the generated electric energy by converting it into hydrogen and synthetic methane can represent an alternative both to the expansion of electricity transmission networks and the electricity distribution networks. Again, it should also be noted that a large-scale expansion of electricity networks—and related significant topographic interventions—in the European states will be accompanied by the resistance of the population (as current much smaller projects already show) with very difficult processes that will considerably hinder the development.

5.1.3 Power-to-Gas as Important Component in Constructing Hybrid Grids

Power-to-Gas can also be a key component to the development of hybrid networks. A hybrid network is hereby a system which arises out of different energy systems that are coupled bi-directionally which is strongly connected and integrated. The implementation of hybrid networks is from an energy system based view of great importance both from the perspective of security of supply as well as from an economic perspective for the future of the European and in particular the Central European energy system. In the area of supply security hybrid networks can provide improved load management as well as energy cross-storages of generation from other networks. From an economic perspective, especially the components of increasing resource efficiency and the reduction of the intensity of a singular expansion of the network and thus the reduction of infrastructure development have to be mentioned. The implementation of hybrid networks enables an optimized integration of existing infrastructure with the involvement of all energy networks in the future—electricity, gas systems, heat power, water supply, transport network. On this basis, strategic decisions can also be made in the energy space planning, making it possible to continually develop the energy system in a regional context and thus also in a supra-regional context that provides crucial positive contributions, so that a strengthening of the domestic economy and living room can be forced.

Heat, electricity and gas grids are nowadays optimized for performance and often run in parallel. Currently, there are links between the electricity and heat power in the form of combined heat and power-CHP technologies, from the power to the heat network by means of heat pumps from the gas to the power grid and district heating supply by power plants and fuel cells. In recent years, due to the development of information and communication (ICT) technologies intersections have increasingly becoming the focus of attention and allow completely new possibilities to arise, which allow for closer coupling of networks, and thus create opportunities in those areas where formerly stand-alone networks were pushed to their limits.

That's why so-called hybrid networks are considered (Lehnhoff 2013). Under a hybrid network (new) interface technologies strongly connect/integrate power systems of different energy networks (e.g. electricity, gas, heat) that are coupled bi-directionally is meant. Hybrid networks offer a great potential for storage capacity and load displacement (Begluk et al. 2013). Highly volatile energy sources such as wind power as well as photovoltaic can be integrated with hybrid networks efficiently and optimally in the energy system. The links between the networks do not only influence the transport of energy sources but enable new forms of storage.

The research necessary for such a network interconnection are partly long-established (e.g. combined heat and power) or are currently being heavily researched (e.g. Power-to-Gas) (Gerhardt et al. 2011). The Power-to-Gas technology allows the connection from the current electricity network to the gas network and therefore the implementation of a full hybrid network (except for the link from the heat to the gas network).

These considerations of interactive networks through steadily growing generation capacity of a fluctuating renewable energy yield such as that from wind and solar, and the additional integration of generation capacity in the power system is also increasing demands on the transmission and distribution networks and in particular the integrated storage capacities. In Austria about 60 % of the demand for electrical energy is already covered by hydropower. However, 100 % coverage would require more renewable energy sources and the requisite massive expansion of energy storage due to the fluctuating generation capacities from wind and photovoltaic sources. The existing potential for pumped storage power plants would need at least a 4–5-fold increase (TU Wien 2011).

Energy storage has a decisive importance in a power system with a high share of renewable energy sources (VDE 2009). In addition to short-term storage, long-term storage is needed due to the seasonal characteristics of photovoltaic and wind (VDE 2012). The electric power system does not meet this storage potential.

In general, a hybrid network with bidirectional coupling possibilities offers, compared to the power grid alone, significantly larger and more time variant storage options. Due to the overall consideration economic (avoid redundancies and unnecessary expansion) and ecological (integration of surplus power) improvements can be achieved.

The implementation of hybrid networks is crucial from an energy system based view both from the perspective of security of supply as well as from an economic perspective for the future of Europe and in particular the Central European energy system. In the area of security of supply hybrid networks can provide improved load management and energy cross-storage. Thus, the realization of hybrid networks allows optimized integration of existing infrastructure with the involvement of all energy networks: electricity, gas systems, heat power, water supply, and the transport sector. Based on this overall perspective, strategic decisions are made in site planning, making the energy system development a priority in a regional context, and also providing crucial positive contributions so that an improvement of the relevant economic and living space can be afforded.

Through coupling technologies or through the realization of hybrid networks today the transfer between energy sources and thus between the networks are not only "smart" and over ICT, they are also possible in both directions. This is an entanglement or improvement of energy transport and network planning if it is economically sound and technically feasible.

The aim of the network-wide optimization is the efficient use of existing infrastructure and energy resources, which means that integration should be planned for in new construction. These new ways of overcoming the problems of individual networks (e.g. capacity limits) not only lead to increased primary energy efficiency (e.g. use of non-integrated and therefore switched off wind energy, avoiding memory leaks) but also increase due to the improved use of existing infrastructures (higher annual utilization), improves profitability.

It will however be necessary that all actors that are faced with new challenges not only cooperate. They also must be able to form the networks inclusive and sustainable by appropriate simulations and subsequent reviews (first economically, second social = cost efficient, thirdly ecologically = resource, environmental and climate-friendly). It then will be possible that in addition to the company, the economy and the climate benefit. For this purpose, not only technical but also socio-economic research is needed.

A non-negligible dimension that will help to realize the Power-to-Gas technology is an increase in the primary energy efficiency in certain sub-segments of the energy system. This is difficult to be stated, since this is highly dependent on the specific application. In the production of hydrogen for use in mobility, with no benefit as a storage option, for example, the overall efficiency of the Power-to-Gas process has to be compared to that of conventional fuels. Here, the efficiency of the power production is of critical importance. All cases produce increased resource efficiency in the economy. In addition, it can be stated that a renewable fuel such as hydrogen or synthetic methane from electrical energy out of wind power, photovoltaic, geothermal or hydro power, in contrast to first-generation biofuels does not face any competition for resources to alternative uses of biogenic resources such as food production. In this context, an increase in the efficiency of resource use would be realized, if the Power-to-Gas product replaces alternative biogenic fuels of the first generation in the mobility sector.

5.2 Several Process Chains

The presentation of the macroeconomic and systemic relevance of Power-to-Gas already shows that Power-to-Gas should not be restricted exclusively on the storage option, it can be useful for other system functions, such as the transportation of energy. Regarding the European energy system, but also quite generally the structure of a power system itself, it becomes clear that Power-to-Gas can still offer more utility functions. The variety of technical options within the Power-to-Gas system shows an extremely wide range of specific embodiments and technological characteristics that will also prevail in different ways in the future energy market. In addition, the required application in the energy system will determine the trends in technological expression. In this chapter, an illustration of the breadth of possible applications of the Power-to-Gas technology and therefore the different process chains takes place.

At the beginning it was mentioned that there is a difference between the intention of the priority development of the Power-to-Gas technology for energy storage and other applications that arise on the basis of energy storage options. The alternative applications are also fundamental to the possibility of storing electrical energy. In particular, long-term storage can build on other specific benefits for the energy system or for certain market participants.

The Power-to-Gas system includes in its broadest sense all technologies and processes in which hydrogen from electrical energy is produced and optionally including carbon dioxide, resulting in the generation of methane. The conversion is limited in the term 'Power-to-Gas' to the production of hydrogen and methane—other transforming forms of electrical energy to hydrocarbons (such as to methanol) are more likely to associate the term 'Power to Fuel' or 'Power to Liquid'. An international standard definition is to the knowledge of the authors still not available. The classification of Power-to-Gas in this context is of central importance, since only in this way a systemic assessment can be made.

Two general applications or processes of Power-to-Gas systems may be differentiated, in the below numeration they are shown:

1. Out of electric energy, which is used for the electrolysis of hydrogen, and out of carbon dioxide, synthetic methane is produced. For this technology, electrical energy is used from renewable energy sources, mainly (but not necessarily exclusively) produced in excess hours from wind power and photovoltaic and stored afterwards in the form of methane. The conversion of hydrogen (H_2) and carbon dioxide (CO_2) into methane (CH_4) is carried out in specially designed facilities.
2. As Power-to-Gas can also be called a system that exclusively produces hydrogen from electrical energy. Hydrogen can also be stored and used directly especially in the transport segment. In addition, hydrogen can be added to natural gas (currently up to 4 % in Austria) so that an application of hydrogen is possible in all energy segments (heat, electricity, transport) (Tichler 2013).

This definition thus includes the exact configuration of the technology of the Power-to-Gas processes or systems which allow for different applications in the energy system. In general, the concept of Power-to-Gas can be described as a very flexible system in terms of the variety of applications and various forms of use. The key technology in Power-to-Gas mixture remains in all applications electrolysis, although various application forms require additional electrolysis forms.

Generally, five identifiable benefits for the energy system can be stated under the definition and in turn different solution strategies for different applications can be understood (Tichler and Gahleitner 2012):

1. To provide a long-term electrical energy storage and the associated improved management of a highly volatile electricity production.
2. The shift of the energy-transport from the power system to the gas system and the associated lower intensity of the expansion of power infrastructure.

3. The ability to raise the share of renewable energy in the transport sector through the use of synthetic methane (but also of hydrogen) from renewable sources (Gahleitner and Lindorfer 2013).
4. The creation of self-sufficient energy solutions in topographically difficult and remote regions for all relevant energy segments: electricity, heat and transport.
5. The use of carbon dioxide as a raw material (and the resulting possible reduction of emission certificates) and the resulting increased resource utilization.

The list of basic capabilities implies in consequence of their various forms of process chains and also different business models with different technology based forms but also with different benchmarks in the energy system. This makes for a compact analysis of the current and expected business forms in terms of the compatibility of the system or of the competitive technologies. As a consequence, an economic evaluation of a specific application of Power-to-Gas system and the associated competing systems or alternative solutions is necessary.

In the following a multitude of possible applications of Power-to-Gas plants is presented. These applications include no rating for economic viability, legal implementation possibility or even in terms of technological expression. The applications are written in a way so that each point has a specific intention for a market participant for the construction and operation of a Power-to-Gas plant. The applications thus relate to a concrete specific benefit for a particular market participant, that can obtained from Power-to-Gas. There is no analysis of the expression and the optimal operation of the plant for a particular business model based on it.

Various applications of a Power-to-Gas plant for the implementation of a particular benefit to market participants of a specific energy market:

I.   A power grid operator implements a Power-to-Gas plant to substitute replacement investments in the electricity grid extension for transmission networks, which he needs for increasing energy transport volumes between supply and demand centers otherwise. Thus, the transport of energy can be transferred to the gas system. The priority intention of establishing a Power-to-Gas plant is thus in the reduction of infrastructure costs in the electricity network.

II.   A power grid operator implements a Power-to-Gas plant in combination with a technology for reconversion—like a fuel cell—for private households, enterprises or technical systems in topographically remote regions to substitute replacement investment in an expensive power grid connection and guarantee a year-round supply. The priority intention of establishing a Power-to-Gas plant is also a reduction in infrastructure costs in the electricity network.

III.   A power grid operator implements a Power-to-Gas plant to solve the load management problem of the power system (especially in the distribution system level) in times of high production of electrical energy from volatile, regional and renewable energy sources by a (temporary) storage of electrical energy and to optimize the system cost to reduce the current balance.

IV.	A gas network operator implements a Power-to-Gas plant in order to achieve a higher utilization of gas networks by shifting the transport of energy from electricity to gas network. The priority intention of establishing a Power-to-Gas plant here represents the expansion of capacity in the network operation.
V.	A potential hydrogen service provider implements a Power-to-Gas plant in order to achieve a higher utilization of the network by shifting the transport of energy from electricity to hydrogen power—equivalent to the gas network operator.
VI.	A wind and/or photovoltaic system operator implements a Power-to-Gas plant in order for a cessation of priority schemes for renewable energy sources in the electricity market and continues to operate the wind turbine or photovoltaic system at any time by an energy storage and conversion which is carried by the Power-to-Gas plant and subsequent feeding into the natural gas grid. Thus, the overall efficiency of the system and thus the annual full load hours can be increased.
VII.	A wind and/or photovoltaic system operator implements a Power-to-Gas plant to takes advantage of any energy unit for intermediate storage to price-to-sell at optimum times in the electricity market. This can be done to optimize the current sale.
VIII.	A biogas plant operator implements a Power-to-Gas plant to increase the use and retention of carbon dioxide with the production of synthetic methane and to increase the overall efficiency.
IX.	A gas storage operator implements a Power-to-Gas plant to achieve through the additional natural gas production, a higher utilization of gas storage at specific times.
X.	A gas storage operator implements a Power-to-Gas plant in order to offer through the production of a new renewable gas product also a renewable storage product.
XI.	A gas trader implements a Power-to-Gas plant to bring a new additional renewable gas product to market that can be sold.
XII.	An electricity producer/trader implements a Power-to-Gas plant to bring a new additional renewable product to market that can be sold.
XIII.	A fuel producer/trader implements a Power-to-Gas plant to bring a new additional renewable product to market that can be sold.
XIV.	An industrial plant (chemical industry) implements a Power-to-Gas plant to offer a new renewable chemical/material product.
XV.	An electricity producer/trader implements a Power-to-Gas plant to use renewable electricity from topographically remote areas with a high potential of renewable energy sources (e.g. Sahara, Patagonia, …) and to be able to transport to the demand centers (e.g. through gas pipelines).
XVI.	A service station operator implements a Power-to-Gas plant to offer a new renewable hydrogen product and to make the supply of hydrogen independent.
XVII.	An industrial plant with a commitment to CO_2 allowances implements a Power-to-Gas plant to bind the otherwise emitted carbon dioxide in

XVIII. synthetic methane, thereby increasing the efficiency and production capacity of existing resources.

XVIII. An industrial plant with a commitment to CO_2 allowances implements a Power-to-Gas to replace fossil sources with renewable sources, thereby reducing the cost of CO_2 allowances to effect.

XIX. A power producer implements a Power-to-Gas plant to provide an additional negative balancing energy and be able to generate revenue on the balancing market.

XX. A power producer implements a Power-to-Gas plant to provide positive balance energy and thus avoid replacement investments in the shade plants.

XXI. The automotive industry implements a Power-to-Gas plant to reduce CO_2-equivalent-emissions of the fleet and thus comply with the legal requirements as well as to offer new products at the same time.

XXII. The operator of a public transportation fleet (e.g. bus, tram, train) implements a Power-to-Gas plant to reduce CO_2-equivalent-emissions of the fleet and to ensure mobility with renewable energy sources.

XXIII. A private household or a business implements a Power-to-Gas plant (in combination e.g. with a fuel cell) to produce energy completely for their own needs and therefore to install a self-contained system to ensure self-sufficiency and a "status symbol".

XXIV. A private household or a business implements a Power-to-Gas plant to intermediate storage power and optimize the current reference costs in the case of flexible rates.

XXV. A producer of an alternative gaseous energy carrier (biogas, coal gas) operates a Power-to-Gas plant to modify the existing gas quality and thus to allow for feeding into the natural gas grid.

XXVI. Regions with a strong potential topographic impact of an electricity grid expansion (or topographical interventions by large conventional energy storage, such as pumped storage power plants) implement Power-to-Gas plants to meet by shifting the transport of energy to the gas network or the alternative energy storage engagement in landscape and/or to reduce in the settlement areas.

XXVII. The "public sector" operates Power-to-Gas plants to increase the share of renewable energy sources, as well as the overall efficiency of the energy system (by reducing the shutdowns of generating plants).

It will relatively quickly be seen that the Power-to-Gas process allows a variety of applications. Of course, the characteristics of a business compatible with the respective benchmarks are also constituted very differently; this will also be discussed in the following chapter.

As a consequence Power-to-Gas can generally be described as a very flexible system in terms of a variety of applications and different forms in the Austrian and international energy systems. The different business models that can be developed based on the specific capabilities of the system also imply specific and varied cases for different market players. This expression of a multifunctional use of Power-to-Gas in the future energy system has also as a consequence a wide economic impact on the technologies. The real intention of the development of the system Power-to-Gas, however, stems from the challenge of a rising generation of volatile generation sources and the necessary option of an additional energy storage that allows for long-term storage.

5.2.1 Systemic Advantages and Disadvantages of Hydrogen and Methane Process Chains

The variety of possible process chains within the system Power-to-Gas shows that as a consequence of specific applications plants focused on the production of hydrogen without methanation as well as plants including methanation will be necessary. Based on conducted analyses inter alia Tichler (2014) a general survey of relative advantages and disadvantages of the energy sources hydrogen and methane out of Power-to-Gas systems in direct comparison can be performed. A respective allotment of certain components to hydrogen or synthetic methane will be attempted due to a positive allocation. Of course every component is relevant for both energy sources:

Advantages of hydrogen compared to synthetic methane from Power-to-Gas-Systems (Reiter et al. 2014):

- Hydrogen shows lower production costs as synthetic methane.
- The production of hydrogen allows a more dynamic driving of the process without additional buffer modules in total.
- The production of hydrogen includes lower conversion losses and therefore the efficiency is better compared to the production of synthetic methane.
- Self-sufficient systems for energy storage can be realised more easily with hydrogen than synthetic methane.
- There is no need of carbon dioxide source for hydrogen production and therefore less local dependence.
- The combustion of hydrogen causes hardly any emissions compared to methane where emissions are released directly.

Advantages of synthetic methane compared to hydrogen from Power-to-Gas-Systems (Reiter et al. 2014):

- The storage of synthetic methane is far less demanding than direct storage of hydrogen—direct storage of hydrogen is more technologically complicated and expensive.
- The usage or transport of synthetic methane can fall back on an existing infrastructure, whereas very few pure hydrogen networks existing today.
- The production and usage of synthetic methane has low restrictions to consumer within the meaning of technology compatibility and in the issue of guarantees compared to hydrogen fed natural gas networks.
- The billing of synthetic methane is far less complex because of similarity or equal calorific value as conventional natural gas—there are minor changes in billing systems in relation to higher hydrogen proportions in natural gas.

- The fed in location into the natural gas network—for storage and transport of energy—is much more problematic in case of fed in, because exact mixing must be ensured—synthetic methane can be fed in without these problems, as long as norms are being met.
- The production of synthetic methane solves the problem definition of an potential dependence on other market participant, which have already admixed the maximum hydrogen amount to the gas distribution system—that problem does not exist in case of synthetic methane.

A total evaluation in terms of determination of a clear advantage can finally only be made on an individual case basis. The sum of key deciding factors must be assessed case-by-case and analysed concerning to the advantages and disadvantages of hydrogen or synthetic methane. A general assessment from the point of view of the authors is not admissible. The flexible Power-to-Gas system allows the use of both energy sources.

5.3 Business Models for Power-to-Gas-Systems

A central question in the appreciation of the economic relevance of a system or a technology is the question whether beside a systemic benefit also an economically positive outcome is present or not; if not, how what justifies an intervention on the part of public institutions. The central question for the framework set to implement new technologies is thus whether in the long-term profitability is given both in business and in the economic context through the launch of a specific product or system. For the market, introduction and development of a financial rate of return is not necessarily meaningful as long as the economic positive relevance of this system is present.

Of course, the realization of market penetration is greatly accelerated with a positive financial rate. The Power-to-Gas system attempts to stress this issue. It has been illustrated that the Power-to-Gas system generates a set of parameters that have a benefit for the energy system and for this reason also for the economy. Of crucial economic importance is also the direct economic effectiveness of the investments involved. Investment in domestic technology components increases the domestic value added.

The Power-to-Gas technology and systems are at the moment at the beginning of their development (single pilot and demonstration plants have been designed or realized in different sizes). The timing of the development stage of the technology may involve no financial rate of return also due to the economic theory. Because of learning curve effects and economies of scale, the cost of production of new technologies in generally decreasing; this is not a special characteristic of Power-to-Gas systems. Here, of course, the speed of implementation and the likelihood of implementation of various forms of system is of immanent importance. The rapid changes in the energy system bring a major challenge with it. Highly innovative products and services—such as new forms of energy storage—should be adequately provided, with which the many new requirements such as the improved integration of energy from renewable sources can be met.

From the perspective of different scientific disciplines, the results of the empirical diffusion research show that the successful diffusion of innovations usually follows an S-curve in which the cumulative adoption component is modeled as increasing over the time function, which is initially convex and becomes a concave function (Rogers 2003). In this context a concrete adoption of these gradients on the Power-to-Gas technology is difficult, since the individual system components are in different stages of development. Thus, the development of dynamic electrolysis tends to be more advanced than those of methane reaction for the production of synthetic methane. Based on the preferred innovation design, which later will prevail in other markets, technological learning can run in a fast speed and therefore the international diffusion can be accelerated (Beise 2001). This in turn requires the knowledge of all future developments and trends in all global markets.

One size that cannot be influenced is certain demand trends for energy storage technologies or energy transport technologies in other regions. This can lead to significant learning curves and economies of scale that can cause decisive cost reductions from technology components also in the Central European market. A crucial influencing factor may be represented by the future demand for back-up systems in regions with poorly developed electricity grid infrastructure.

Currently, the demand for Power-to-Gas systems is a priority in Central Europe; occasionally this is also the case in other regions (such as in Canada or in French overseas colonies). The global demand, however, could be significantly determined by back-up systems for decentralized systems in developing and emerging countries with suboptimal or poor network expansion degree. In India is due to telecom operators a high number of inadequately supplied with electrical energy Telekom plants that are currently operating with inefficient and expensive diesel fuel generators. Back-up systems with electrolysis units and fuel cells that have integrated hydrogen generation and storage and a reconversion, can provide a crucial and promising alternative. The high number of regions with insufficient supply due to poorly developed electricity grid infrastructure, particularly in Asia but also in Africa and South America could thus generate a large demand for power back-up systems with dynamic electrolysis units, whereby the whole production of the electrolyzer can be raised to a semi or fully automated stage. The investment costs of the dynamic electrolyzer are fundamentally influenced by the increase in technology, so that the learning curve would significantly change in production at larger scales. At the moment an accurate assessment of this development is not possible.

The following forms an approximate estimate of the cost characteristics of Power-to-Gas plants, with reference to the large range of technological, legal and also economic differences. It is explicitly pointed out that for the evaluation of the economic manifestation of the Power-to-Gas system, the deployment and use of technology is important and therefore the respective specific markets are developed. Thus, the economic analysis of the different business models of a

Power-to-Gas plant for a position in the overall system must always be interpreted in terms of their systemic benchmarks. Of course, the characteristics of a business competition compatibility with the respective benchmarks are also constituted very differently. This makes for a compact analysis of the current and expected business forms in terms of the compatibility of the system or of the competitive technologies. As a consequence of an evaluation of a specific application of the Power-to-Gas system and the associated systems competition or alternative solutions are separately necessary.

In this context, the present book cannot go into detail on the different cost characteristics. For a more detailed examination and analysis of current and future costs and cost components see (Steinmüller et al. 2014).

Evaluations of different process chains generally show that the flexible energy storage instrument Power-to-Gas brings very different production costs with it depending on the specific application. The quantitative economic analysis in Steinmüller et al. (2014) illustrates that the current investment cost of Power-to-Gas plants are relatively high. The current technologies—both hydrogen production based on dynamic electrolysis and the subsequent methanation are still being developed.

The largest share of the investment costs of a Power-to-Gas plant is clearly in the process of electrolysis, followed by the methanation plant component. In the future development of investment costs, the learning curves and economies of scale must therefore be taken into consideration especially for these two components. In the future reduction of the investment costs of electrolysis and methanation the technological learning curve plays a large role. A distinction is made between the cost reduction by improving the technology and the reduction by increasing the cumulative capacity (greater number and size of installed systems). According to Grond et al. (2013) for alkaline electrolysis an annual reduction of costs through improved technology of about 0.4 % is expected. In PEM electrolysis the improvement potential of 2.2 % is assessed denoting a significant annual increase. The study from (Schoots et al. 2008) determined learning rate for electrolysis is kept at 18 % and this was confirmed in Steinmüller et al. (2014) by the analyzed values. The annual cost reduction potential through technological improvement of methanation is in Steinmüller et al. (2014) with 2 % pa analyzed. A cost reduction is achieved at methanation plants, unlike electrolyzers, only at higher system capacities.

As a consequence, in the current technology stage, the total production costs of hydrogen and/or synthetic methane are highly dependent on the achievable full-load hours. Another key factor beside the achievable full-load hours is the rated power of the electrolyzer used since the specific investment costs decrease with increasing power. These economies of scale occur equivalently also at the methanation reactor. A further development of the technology components with an associated cost reduction is essential for a business use of Power-to-Gas systems.

The final cost of methane from Power-to-Gas plants is due to the increased investment demand and lower efficiency from the additional process step of methanation in all process chains generally higher than those of hydrogen.

Nevertheless, in certain applications a methanation can be necessary and useful, for example if the hydrogen that is fed into the natural gas network is not useable.

According Koppe (2014) the indication by the current, specific costs can only be made with approximate standard values because the actual cost of the respective plant equipment depends on the intended use of the system. Today, for the pure electrolysis stack investment costs of €1,000 (Koppe 2014) for AEC electrolyzers achieved with PEM stacks around €2,000. Depending on the size of the installed capacity and the technology of electrolysis, the investment cost per kW of installed electrical power changes. Including all peripheral electrolysis current costs in 2014 are between 1,500 €/kWe and 9,000 €/kWe. For the year 2030, Steinmüller et al. (2014) predicted that for both alkaline and PEM electrolysis including electrolyzer peripherals (stack + O) costs of less than 1,000 €/kWe for larger systems. For the methanation in the form of methane reactions costs for comparison from Lehner (2014) are for larger systems of future investment 130–300 €/kWe.

Of central relevance to the economic viability of Power-to-Gas installations are the investment costs; particularly the development of pricing structures and the so-called price spreads on spot markets for electricity. It is about the difference between the prices at which electricity can be purchased in comparison to the prices in the current sale. If there are not very large Power-to-Gas plants in focus, electricity costs, electricity demand and the efficiency of the electrolyzer and the specific carbon dioxide costs for methanation have in relation to the investment costs (especially at low full-load hours) secondary importance. Potential additional revenues like waste heat or oxygen proceeds are to be described in this context as secondary. This has been investigated in various individual process chains in Steinmüller et al. (2014). The final cost can be reduced only slightly by a byproduct of electrolysis; the sale of waste heat from the methanation.

A comprehensive assessment of the economic production costs is only in comparison with the respective specific benchmark technologies entirely possible. Again it should be emphasized that the different Power-to-Gas process chains and the different business models have different intentions, products, sizes and applications and the resulting production costs differ greatly.

If Power-to-Gas is used for storing electrical energy, pumped storage and compressed air storage are in this case the specific benchmark technologies (further options are given by an extension of the transmission grids or demand-side reactions). In this process chain, the priority use of surplus electricity from the power grid to produce hydrogen or methane in a Power-to-Gas plant is taken as a base. The produced gas shall then be fed into the public natural gas grid and be available for various applications. Power-to-Gas plants that use excess electricity for the production and supply of hydrogen and synthetic methane should be used in areas with a high share of fluctuating electricity generators. With a small but high demand for electricity production from renewable generation technologies, such as a wind farm, these investments would otherwise be turned off. The generated hydrogen or synthetic methane is thus fed into the existing local natural gas grid. The decisive factors are the maximum allowable volume fraction of hydrogen in natural gas as well as the bottleneck capacity at each location.

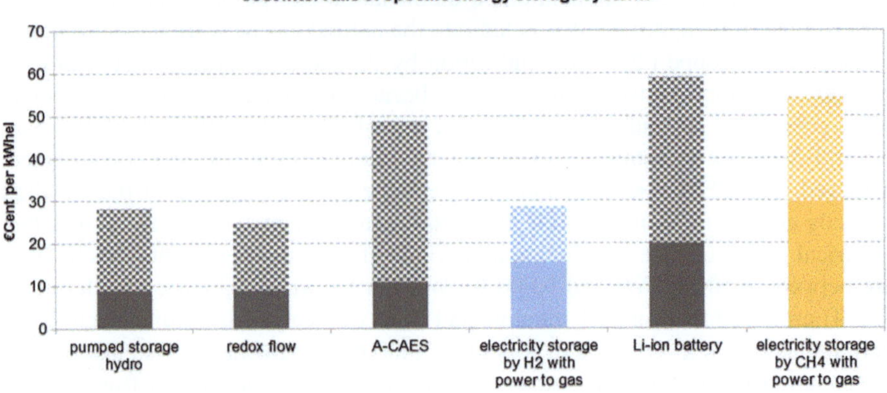

Fig. 5.2 Comparison of intervals of specific energy storage systems (with an enhanced Power-to-Gas system). *Source* own figure, based on data of (Steinmüller et al. 2014). *Notes* (1) no combinations of different business models integrated. (2) Power-to-Gas integrates enhancements of the Power-to-Gas technology by realizing both learning curves and scale effects. (3) Power-to-Gas includes herby also an additional re-converting of hydrogen respectively methane to electricity—because the listed alternative solutions always imply a storage of electricity without alternatives. (4) Power-to-Gas costs in the figure are shown for the actual Austrian legislative framework (no electricity grid tariffs, no additional fee)

Figure 5.2 illustrates the cost characteristics of the Power-to-Gas system with an exclusive focus on electricity storage in relation to the current cost of alternative electricity storage solutions. The costs of Power-to-Gas includes herby also an additional re-converting of hydrogen respectively methane to electricity for a direct comparison of an electricity storage. The costs integrate enhancements of the Power-to-Gas technology by realizing both learning curves and scale effects. For a detailed composition of cost components of these Power-to-Gas plants see Steinmüller et al. (2014).

The difference in the cost of production is not significant in any case but the technologies differ, however, in terms of efficiency and expansion potential. The efficiency along the entire process chain of Power-to-Gas is significantly lower, the potential for expansion of pumped storage and compressed air energy storage is limited and these technologies are highly location-dependent, also no long-term storage is possible among the alternatives to Power-to-Gas.

As illustrated, hydrogen and methane from Power-to-Gas plants can be used in transportation. As benchmarks both biogenic and fossil fuels can also be used. In this Power-to-Gas process chain electricity from the public power grid is used to generate a renewable product for the transportation sector. This may be some part hydrogen for use in fuel cell vehicles, and some part methane for use in CNG vehicles. The transport of the product gas from the Power-to-Gas installation can be done through the natural gas network for hydrogen transportation but can use hydrogen pipeline or trucks with pressure tanks. The volume of purchased electricity in this process chain is not tied exclusively to the excess quantities in the

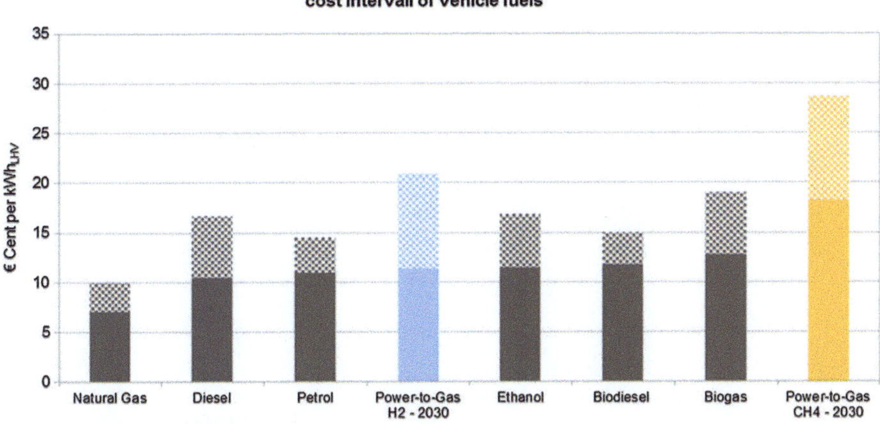

Fig. 5.3 Comparison of intervals of specific vehicle fuels (with an enhanced Power-to-Gas-system). *Source* own figure, based on data of Steinmüller et al. (2014). *Notes* (1) no combinations of different business models integrated. (2) Power-to-Gas integrates enhancements of the Power-to-Gas technology by realizing both learning curves and scale effects. (3) Power-to-Gas costs in the figure are shown for the actual Austrian legislative framework (no electricity grid tariffs, no additional fee)

public power grid and thus Power-to-Gas plants can be used with larger power ratings.

Figure 5.3 illustrates the cost characteristics of the Power-to-Gas system with an exclusive focus on producing a vehicle fuel in relation to the current cost of alternative fuel costs. The costs of Power-to-Gas includes herby an optimization of the electricity input, naturally without re-converting of hydrogen respectively methane to electricity. The costs integrate enhancements of the Power-to-Gas technology by realizing both learning curves and scale effects. For a detailed composition of cost components of these Power-to-Gas plants see Steinmüller et al. (2014).

The comparison of the production costs show that they are currently higher still for H_2 or CH_4 from Power-to-Gas, but in the future this process can compete with conventional fuels, depending on future price developments. Differences arise in the compared drive concepts and fuels especially in the greenhouse gas emissions and the amount of electrical potential and space requirements. Through a future cost reduction in the technology Power-to-Gas may well compete with the remaining fuel costs from an economic perspective.

In another process chain, the Power-to-Gas plant can make in combination with a fuel cell, a self-sufficient system, for example, for a topographically remote area. In this case hydrogen from the electrolysis is used as the storage medium for electric energy. Electricity from a photovoltaic system in an autonomous system (stand-alone solution) is used in an electrolyzer to produce hydrogen. The dimensions of the individual components depend strongly on the location and size of each stand-alone solution. The hydrogen is then stored locally and converted back into a fuel cell when needed. There is no connection to the public electricity or gas network.

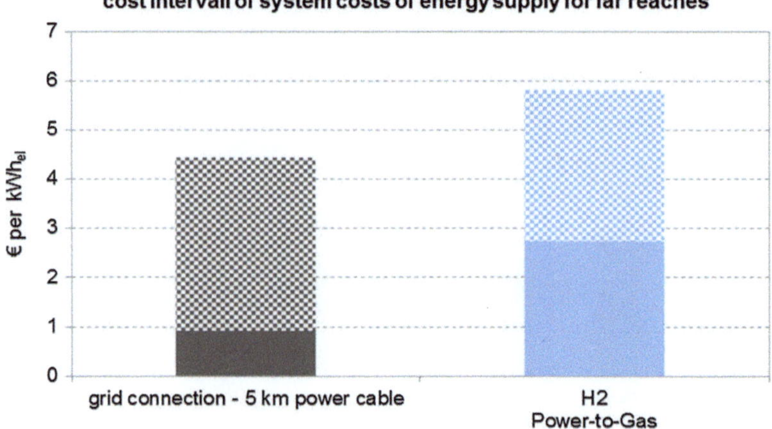

Fig. 5.4 Comparison of intervals of different system costs of energy supply for far reaches (with an enhanced Power-to-Gas-system). *Source* own figure, based on data of (Steinmüller et al. 2014). *Notes* (1) no combinations of different business models integrated. (2) Power-to-Gas integrates enhancements of the Power-to-Gas technology by realizing both learning curves and scale effects. (3) Power-to-Gas costs in the figure are shown for the actual Austrian legislative framework no electricity grid tariffs, no additional fee). (4) Power-to-Gas costs include costs for a fuel cell for re-converting the hydrogen to electricity and a small hydrogen storage system

The Fig. 5.4 illustrates the cost characteristics of the Power-to-Gas system with an exclusive focus on a stand-alone system with hydrogen in relation to the current costs of an energy supply for far reaches. The costs of Power-to-Gas includes herby a direct usage of electricity out of an own photovoltaic module, a small hydrogen storage system and costs of re-converting the hydrogen to electricity with a fuel cell. The costs integrate enhancements of the Power-to-Gas technology by realizing both learning curves and scale effects. For a detailed composition of cost components of these Power-to-Gas plants see Steinmüller et al. (2014).

The high production costs of caring for a stand-alone system with hydrogen can be put into perspective by comparing it with the cost of a power outlet in a remote area. Although—as in the Fig. 5.4—a connection of a single building with low annual electricity requirements and a wide distance of 5 km to the main power source is certainly an extreme example, though there are in reality certain examples where these conditions apply. Despite the high costs of Power-to-Gas, its use for powering autonomous systems can be an economically viable application.

Another possibility for the use of Power-to-Gas is also the transport of renewable energy generated in remote regions in the demand centers. In this process chain, the natural gas network should be used to transport renewable energy from remote areas in the demand centers. Remote regions often have a high potential of renewable energy sources (wind and solar), but usually do not have electricity demand. In order to transport the electricity generated from renewable sources in the demand centers, it is converted into hydrogen or methane in a Power-to-Gas plant and fed into the natural gas network. There, the gas is then available for different applications. Alternatively, the current could also be transported with an HVDC line (high voltage direct current transmission), but the building of new energy infrastructure would then be necessary.

Examples of high potential for renewable energy sources are the large wind potential at sea or in coastal regions or the high insolation in desert areas. In these remote areas, however, no need for electrical energy is usual and a transport option is required to the demand centers. The energy transport can be realized with Power-to-Gas by the generated current to produce hydrogen or methane. This energy can be fed into the natural gas grid and transported to regions with higher energy requirements. An alternative to Power-to-Gas are HVDC lines with which the generated electrical power can be transported directly.

Areas of application for this process chain arise primarily in regions with very high potential for renewable energy sources such as wind or solar energy. In order to transport the generated energy in the demand centers, an access to the natural gas network is needed. If methane should be generated in the Power-to-Gas plant, a carbon dioxide source is required. This could be a challenge in remote areas and may require transport of carbon dioxide. The waste heat from the methanation could be used in this process chain either for CO_2 capture or preheating in a solar thermal power plant.

The costs of the transport of energy by means of HVDC transmission from a remote area with high renewable potential to the demand centers currently are below the cost of using the Power-to-Gas technology. Here, however, it should be noted that no transport distance is set and therefore no transport losses were taken into account. Although the efficiency of Power-to-Gas by the additional conversion to an energy carrier is lower the costs are higher and this could address other aspects of the use of this technology. One of these aspects is, for example, the low acceptance for network infrastructure projects in the population. Through the feeding of hydrogen or methane into the natural gas grid the Power-to-Gas technology would use largely existing infrastructure (Fig. 5.5).

Overall, analyses show that currently a Power-to-Gas plant (H_2 and CH_4) is economically far from being competitive with alternatives such as conventional fed biogas. However, it is clearly pointed out that Power-to-Gas systems are a current technology under development. In general, learning curve effects and economies of scale reduce the costs associated with new technologies. In addition, it should be noted that current calculations include very low power levels, whereby the dominance of high investment costs strongly influence (in relation to low cost of ownership) the result.

In general it can be stated that current issue of the economic viability of the Power-to-Gas operation is significantly overshadowed by the unique positive system benefits of the technology. For economic and welfare benefits to be realized Power-to-Gas as a solution option, has to be pursued and supported from the public sector.

In addition, approximate analyses show that even for higher-level macroeconomic effects (based on the cost characteristics) carried out in (Steinmüller et al. 2014), the construction of Power-to-Gas plants have positive effects on the

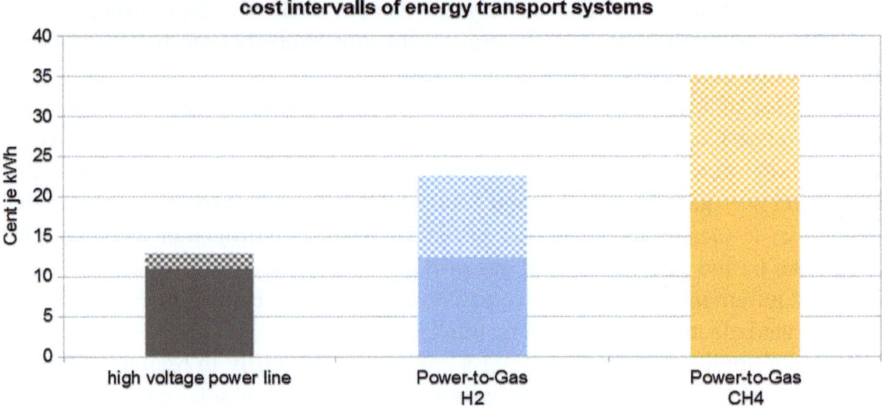

Fig. 5.5 Comparison of intervals of specific energy transport systems for integrating electricity surpluses out of volatile production into the system (with an enhanced Power-to-Gas-system). *Source* own figure, based on data of Steinmüller et al. (2014). *Notes* (1) no combinations of different business models integrated. (2) Power-to-Gas integrates enhancements of the Power-to-Gas technology by realizing both learning curves and scale effects. (3) Power-to-Gas costs in the figure are shown for the actual Austrian legislative framework (no electricity grid tariffs, no additional fee)

central European economies. It has to be stated that the realization and operation of Power-to-Gas plants leads to a higher gross domestic product as well as higher levels of employment. The business model should be based on a higher production and alternative energy imports are substituted by energy storage. A potentially higher gross domestic product as well as positive effects on employment through the implementation of Power-to-Gas plants is based on 5 positive and 1 negative aspects, compared to a situation without implementation of the systems.

As a positive component of the construction of Power-to-Gas plants it has to be mentioned that the investment momentum may generate additional spending from companies for the construction of Power-to-Gas plants and thus additionally generate value outside the energy sector in the production of technologies as well as in the construction sector. In addition, the substitution of imported energy through domestic energy production in the form of hydrogen or synthetic methane from additional wind turbines or photovoltaic systems causes a significant value-added profit. This is based on the increased use of electric energy from generating plants by volatile energy storage in times of highly volatile production. Positive consequences through the knowledge transfer as well as the use of domestic technology components and a consequent production and export of technology have to be completed. All investment increases also produce positive effects on the employment (especially in the construction industry and in the technology of production), this in turn leads to increased payroll and then triggers higher private consumption. However, it should also be stated that the circulation of the higher energy prices because of the higher costs of production in the current technology stage generates partially negative effects to the end users.

The greatest pronounced influence on a positive macro-economic result is the substitution of imported energy through domestic energy production, thereby resulting in significant value-added profit and a significant increase in net exports. This is fundamentally supported by the positive investment impulses, of which construction, manufacturing and service companies benefit. In Central Europe energy imports will be substituted. A qualitative assessment of the macroeconomic impacts of the construction and operation of Power-to-Gas plants in Central Europe shows that triggered multi-round effects are generally positive, although the financial burden has also multiple partially negative effects on earnings.

5.4   Legal Aspects

A general analysis of Power-to-Gas is—just like any other technology—faced with different national legislations. A comprehensive legal analysis of a specific technology requires ever more the analysis of specific national laws. There are, for example, in Germany different applicable legislations and eligibility conditions as in Austria or France. This complicates the presentation and analysis of legal aspects for Power-to-Gas without specific local or national context.

For this reason the general need to define the legal framework for Power-to-Gas will be discussed in this chapter. In addition, basic problems that come from the legal framework of the European Union are discussed in the context of Power-to-Gas and the need for positive public regulation to implement Power-to-Gas into the energy system.

The exact definition of a specific legal framework is also from an economic and technological point of view of imminent importance. Only the existence of legal certainty allows activities of significant investment, on the one hand to the promotion of research and development and on the other hand to the realization of investments to systemic and business purposes. However, the legislators are faced with the problem that technologies that are at the beginning of their development are not tangible to the market. Therefore the adaptation of the legal framework prior to the implementation of the technology on the market is hard as complications cannot always be predicted. For this reason expanded research in the area of energy law has to be done in the next few years in the context of Power-to-Gas at both national and international level.

A significant problem that the Power-to-Gas system is faced with in Europe is the liberalization of the electricity and gas market. This complicates the implementation of concepts of the intelligent Power-to-Gas systems that also cause a system-wide benefit. With the liberalization only the entire process chain can in isolated cases be realized. For example, if they are operated by the possession of private gas entities. Thus a higher-level system benefits the economy and the energy system at risk because those market participants who do not benefit from the storage and transport technology represent those market participants who have to bear the costs in terms of investment and operating costs.

Furtlehner (2014) explains in Steinmüller et al. (2014) the meaning and purpose of market liberalization within the third internal market package of the European Union:

> The liberalization of energy markets plays a central role for Europe's competitiveness.The European Union has redesigned with the adoption of the third internal market package in 2009, the legal framework for the internal energy market. Key aspects of the third internal market package are stricter rules of unbundling of transmission system operators and transmission companies, and options for ownership unbundling, have Independent System Operators (ISO) and the Independent Transmission Operators (ITO) to choose from. The independence of the transmission system operators and the transmission companies is intended to ensure a range of measures. Other key points of the third internal market package concern consumer protection and energy poverty, the expansion of the powers of the regulatory authority, the introduction of a distance-independent pricing of gas transportation via pipeline networks (entry/exit tariff), and smart metering and the creation of an Agency for Cooperation of Energy Regulators.
>
> The third internal market package consists of the following instruments: Regulation (EC) No 713 2009 establishing an Agency for the Cooperation of Energy Regulatory Commission Regulation (EC) No 714/2009 on access to the network for cross -border exchanges in electricity, Regulation (EC) No 715/2009 on conditions for access to the natural gas transmission networks and Directive 2009/72/EC concerning common rules for the internal market in electricity and Directive 2009/73/EC concerning common rules for the internal market in gas.
>
> The so-called Internal Gas Market Directive provides in Article 1 on the subject matter and scope that, firstly, this Directive establishes common rules for the transmission, distribution, supply and storage of natural gas and the policy, the organization and functioning of the natural gas sector, market access, establishes the criteria and procedures for the granting of transmission, distribution, supply and storage of natural gas and the operation of systems. Second, the rules established by this Directive for natural gas also apply in a non-discriminatory way to biogas and gas from biomass or other types of gas, as far as it is technically possible and without compromising safety, feeding these gases into the gas grid and transported through the network.

The equality of intermediate products of Power-to-Gas, hydrogen and synthetic methane, natural gas and biogas require the acquisition of problems in the wake of new technological and systemic innovative solutions. The liberalization of the electricity market can generally be seen as a problem in the context of networking sectors of heat, transport and electricity through Power-to-Gas, as a new definition of the roles of market participants along the supply chain will be necessary. The unbundling is therefore considered more likely to be counterproductive for this systemic approach. Market liberalization and the previously mentioned problem of the different necessary market participants along the supply chain thus making it difficult in this context, the implementation of the system benefits of storage technologies. For an optimal system an adaptation of the framework, the development of entirely new approaches and a new design of market rules are required. At present, in the field of storage technologies and Power-to-Gas every market participant has to optimize on his own. The system-wide optimization is not given enough regard. The responsibilities and players' roles in the energy system must be redefined for the implementation of new storage technologies and systems.

An additional aspect is from an economic perspective, existing market distortions in different energy systems. Market distortions from an economic perspective in this context are, for example, a lack of internalization of externalities, both negative externalities and positive externalities. The provision of storage technologies by market participants and due to the free-rider issue non-existent compensation of the end user for the storage of energy in the system is a positive external effect. This positive external effect is an additional benefit that is currently inadequately capitalized on most market systems. As long as the public sector does not engage in regulation, market distortion will cause inadequately installed memory technologies in the energy markets.

In addition to the existing technological challenges in Power-to-Gas and the field of energy storage technologies, is a need to address the adaptation of the framework in the energy system, as currently monetary incentives for the realization of storage technologies are lacking. Should it not therefore come to a fundamental system adaptation, then there will be too little storage in the future in order to optimally use the energy system. This reinforces the problem of the long payback periods of storage infrastructures. Despite a systemic point of view, medium to long-term profitability avoid high initial investment also due to the general market uncertainties, the realization of large projects and also of storage infrastructure. It therefore requires the implementation of new and legally anchored solutions, both in the financial market as well as in funding and fiscal systems.

In addition, by introducing the Power-to-Gas technology in different legislation, there is a specific need for action to adapt a legal framework that allows for the installation of Power-to-Gas plants in the first place.

In Germany, for example, the relevant legal standards for the production and supply of storage gases in the German gas network, the Energy Act and the regulations based thereon, in particular the Gas Network Access Ordinance and the Gas Network Charges Ordinance are anchored. Furtlehner (2014) further explains:

> Also of importance for the production and supply is the Renewable Energy Sources Act (EEG). With the Act revising energy economy legislation from 26.7.2011 the German legislature in § 3, paragraph 10c Energy Industry Act (Energy Act), the definition of the term 'biogas' supplements. This now also includes hydrogen, which is produced by electrolysis of water, and synthetically produced methane when the power used for the electrolysis and the CO_2 used for methanation of carbon monoxide or each shown to predominantly come from renewable energy sources [...]. Predominantly is intended according to the explanatory memorandum mean a share of at least 80 %. Hydrogen from renewable and synthetic methane with it - as well as landfill gas, sewage gas and mine gas - simply defined as 'biogas'. A separate legislative framework for these substances is not yet established. The new definition has far-reaching consequences. The term 'biogas' for benefit hydrogen and synthetic methane, provided that the other requirements of § 3, paragraph 10c of the Energy Act are met, from all energy law provisions that favor biogas in relation to other energy sources.

While the adaptation of the legal framework is already very advanced in Germany, there is a need for Austria to adapt the legal framework as Furtlehner (2014) explains:

> While the German legislature in § 3, paragraph 10c Energy Industry Act (Energy Act), the definition of 'added Biogas', found in the Natural Gas Act of 2011, Austria has no comparable definition. An important question now is what this discrepancy in the implementation for the supply of methane, hydrogen and synthetic means in the natural gas

grid. There are many indications that in the present case there is a gap, since the determination of the Natural Gas Act 2011 is incomplete in the light of the Directive and this incompleteness is not intended by the legislature. Currently it cannot be concluded with certainty that there is a legal basis in the Austrian legal system for supplying hydrogen or synthetically produced methane into the gas grid. Should this be accepted, the provisions of ÖVGW RL 31 in any case apply or ÖVGW RL 33.

There is in addition a need for further development of technological components and the economic models as well as a need for further development of the legal framework, both at the national level, as well as superior multilateral level. Thus the positive solution of the Power-to-Gas technology can realize the greatest value for the energy system in the future.

References

Amt der NÖ Landesregierung (2011) NÖ Energiefahrplan 2030. http://www.noe.gv.at/Umwelt/E nergie/Energiezukunft/energiefahrplan.pdf. Accessed 30 May 2014

Begluk S et al (2013) SYMBIOSE und Speicherfähigkeit von dezentralen Hybridsystemen; Presentation at 8. Internationale Energiewirtschaftstagung an der TU Wien, Wien 13–15 Feb 2013

Beise M (2001) Lead markets: country specific success factors of the global diffusion of innovations. ZEW Economic Studies, vol 14, Physica-Verlag, Heidelberg

Brauner G (2003) Blackout Ursachen und Kosten. Energy 4/03

Cohen J, Schmidthaler M, Reichl J (2014) Re-focussing research efforts on the public acceptance of energy infrastructure: a critical review. doi:10.1016/j.energy.2013.12.056

dena—Deutsche Energie-Agentur-GmbH (2010) dena-Netzstudie II. Integration erneuerbarer Energien in die deutsche Stromversorgung im Zeitraum 2015–2020 mit Ausblick 2025

Energieinstitut an der Johannes Kepler Universität Linz GmbH (2012) Technologiekonzept power to gas. Brochure

Furtlehner M (2014) In: Steinmüller H, Tichler R, Reiter G et al (eds) (2014) Power to gas—a systems analysis. Project report for the Austrian Federal Ministry of Science, Research and Economy

Gahleitner G, Lindorfer J (2013) Alternative fuels for mobility and transport: harnessing excess electricity from renewable power sources with power to gas. ECEEE 2013 summer study, France

Gerhardt N, Jentsch M, Pape C, Saint-Drenan Y, Schmid J, Sterner M (2011) Speichertechnologien als Lösungsbaustein einer intelligenten Energieversorgung - Fokus Strom-Gasnetzkopplung. In: Presentation at E-world energy and water 2011—Smart Energy, Essen, 8–10 Feb 2011

Grond L, Schulze P, Holstein J (2013) Systems analysis power to gas: a technology review, Groningen

Hefner R III (2007) The age of energy gases. China's opportunity for global energy leadership. The GHK Company, Oklahoma City

Igel M, Winternheimer S, Fixemer R, Leinenbach J (2010) Netzintegration von Solarstromerzeugung. Teil 2. ew – Das Magazin für die Energie-Wirtschaft 109(6):33–37

Koppe M (2014) In: Steinmüller H, Tichler R, Reiter G et al (eds) (2014) Power to gas—a systems analysis. Project report for the Austrian Federal Ministry of Science, Research and Economy

Lehner M (2014) In: Steinmüller H, Tichler R, Reiter G et al (eds) (2014) Power to gas—a systems analysis. Project report for the Austrian Federal Ministry of Science, Research and Economy

Lehnhoff S (2013) Hybridnetze für Smart Regions. Presentation at the conference Energieinformatik, Wien, 13 Nov 2013

Oesterreichs Energie (2010) Klimaschutz durch modernste Technik. Pumpspeicher bilden Schwerpunkt der Kraftwerksprojekte der E-Wirtschaft. Oesterreichs Energie - Fachmagazin der österreichischen E-Wirtschaft, November/Dezember 2010

Reichl J, Kollmann A, Tichler R, Schneider F (2006) Umsorgte Versorgungssicherheit. Trauner Verlag

Reiter G, Tichler R, Steinmüller H et al (2014) Wirtschaftlichkeit und Systemanalyse von Power-to-Gas-Konzepten. In: DVGW (2014) Technoökonomische Studie von Power to gas Konzepte. Forthcoming

Rogers E (2003) Diffusion of innovations, 5th edn. Free Press, New York

Schmiesing J (2010) Neue Hausausforderungen für ländliche Verteilnetzbetreiber durch dezentrale EEG-Einspeisung. Presented at the conference Aktuelle Fragen zur Entwicklung der Elektrizitätsnetze. Ermittlung des langfristigen Ausbaubedarfs, Göttingen, 15 Apr 2010

Schoots K, Ferioli F, Kramer G, van der Zwaan B (2008) Learning curves for hydrogen production technology: an assessment of observed cost reductions. Int J Hydrogen Energy 33(11):2630–2645

SRU (2010) Sachverständigenrat für Umweltfragen SRU 100 % erneuerbare Stromversorgung bis 2050: klimaverträglich, sicher, bezahlbar. p 479

Steinmüller H, Tichler R, Reiter G, Koppe M, Lehner M, Harasek M, Gawlik W, Haas R, Haider M, et al (2014) Power to gas—A systems analysis. Project report for the Austrian Federal Ministry of Science, Research and Economy (This work was funded and co-funded by the Austrian ministry of economics, by Österreichs Energie and by FGW. This report is only available in German language and can be ordered from Energieinstitut an der Johannes Kepler Universität Linz)

Tichler R (2011a) Der mögliche Beitrag von SolarFuel als neue power to gas-Technologie für eine zukünftige europäische Energieversorgung. In: Steinmüller H, Hauer A, Schneider F (eds) Jahrbuch Energiewirtschaft 2011. Neuer Wissenschaftlicher Verlag

Tichler R (2011b) Analysen zur Weiterverfolgung der power to gas-Technologie. Vergleich derzeitiger Förderungen sowie aktueller Steuersätze in den Segmenten Strom/Wärme/Verkehr zur Kalkulation einer preislichen Gleichstellung eines synthetischen power to gas-Produktes durch Förderungen/Einspeisetarife in Österreich. Energieinstitut an der Johannes Kepler Universität Linz GmbH

Tichler R (2013) Volkswirtschaftliche Relevanz von power to gas für das zukünftige Energiesystem. Presentation at 8. Internationale Energiewirtschaftstagung an der TU Wien, Wien 13–15 Feb 2013

Tichler R (2014) In: Steinmüller H, Tichler R, Reiter G et al (eds) Power to gas—a systems analysis. Project report for the Austrian Federal Ministry of Science, Research and Economy (This work was funded and co-funded by the Austrian ministry of economics, by Österreichs Energie and by FGW. This report is only available in German language and can be ordered from Energieinstitut an der Johannes Kepler Universität Linz)

Tichler R, Gahleitner G (2012) power to gas – Speichertechnologie für das Energiesystem der Zukunft. Energieinstitut an der Johannes Kepler Universität Linz, Energie Info 08/2012

Tichler R, Steinmüller H, Hauer A, Pengg-Bührlen H et al (2011) Machbarkeitsstudie einer SolarFuel β-Anlage in Österreich. Energieinstitut an der Johannes Kepler Universität Linz GmbH, SolarFuel GmbH

TU Wien, ESEA/EA (ed) (2011) Super-4-Micro-Grid - Nachhaltige Energieversorgung im Klimawandel. approbierter Endbericht zum Forschungsprojekt im Rahmen der 1. AS Neue Energien 2020, Wien

VDE (2009) Energiespeicher in Stromversorgungssystemen mit hohem Anteil erneuerbarer Energieträger – Bedeutung. Stand der Technik Handlungsbedarf, Frankfurt

VDE (2012) Energiespeicher für die Energiewende – Speicherungsbedarf und Auswirkungen auf das Übertragungsnetz für Szenarien bis 2050, Frankfurt

MIX
Papier aus verantwortungsvollen Quellen
Paper from responsible sources
FSC® C105338

If you have any concerns about our products,
you can contact us on
ProductSafety@springernature.com

In case Publisher is established outside the EU,
the EU authorized representative is:
**Springer Nature Customer Service Center GmbH
Europaplatz 3, 69115 Heidelberg, Germany**

Printed by Libri Plureos GmbH
in Hamburg, Germany